Water Services Management and Governance

Governance and Management for Sustainable Water Systems Series

Water Services Management and Governance

Lessons for a Sustainable Future

Editors

Tapio S. Katko, Petri S. Juuti, and Klaas Schwartz

Assistant Editor

Riikka P. Rajala

IWA Publishing

London • New York

Published by **IWA Publishing**
Alliance House
12 Caxton Street
London SW1H 0QS, UK
Telephone: +44 (0)20 7654 5500
Fax: +44 (0)20 7654 5555
Email: publications@iwap.co.uk
Web: www.iwapublishing.com

First published 2013

British Library Cataloguing in Publication Data
A CIP catalogue record for this book is available from the British Library

Library of Congress Cataloging-in-Publication Data
A catalog record for this book is available from the Library of Congress

ISBN: 9781780400228 (Paperback)
ISBN: 9781780400730 (eBook)

Contents

About the Authors

Editors

Dr. Petri S. Juuti is a historian, and Adjunct Professor in the universities of Tampere, Oulu and Turku. He is currently working as university researcher at the School of Social Sciences and Humanities University of Tampere. His major area of interest is the urban environment, especially city service development, water history, urban technology, pollution, and public policy. Author and editor of over 30 books and more than 30 refereed articles. petri.juuti@uta.fi

Dr. Tapio S. Katko is a civil engineer, Adjunct Professor, and UNESCO Chairholder in Sustainable Water Services at Tampere University of Technology. His main research interests are long-term development of water and related infrastructure as well as institutional, management, policy and governance issues of water services. Author of several books. Coordinating editor. tapio.katko@tut.fi

Dr. Klaas Schwartz is Senior Lecturer Urban Water Governance in the Department of Integrated Water Systems and Governance at the UNESCO-IHE Institute for Water Education. His main area of interests concerns public sector reform, informal water provisioning and partnerships in service provision. He has been involved in research, consultancy and training activities in the field of Urban Water Governance in various countries in Asia, Africa and Latin America. k.schwartz@unesco-ihe.org

Assistant editor

Dr. Riikka P. Rajala (Environmental Engineer) is currently working as a Post-Doctoral researcher in University of Tampere. Her areas of expertise and research interest are the environmental history and history of technology, urban history, sustainable water and sanitation services and water services management. Rajala is the author or co-author of over 12 scientific monographs and more than ten refereed articles. riikka.rajala@uta.fi

Chapters

Ch.1.
Jeltsje Kemerink, Lecturer Integrated Water Resources Management, Department of Integrated Water Systems and Governance, UNESCO-IHE Institute for Water Education, j.kemerink@unesco-ihe.org
Mbuvi Dorcas, PhD candidate, UNU-MERIT, Maastricht University, dorcas.mbuvi@maastrichtuniversity.nl

Ch.2.
Kimmo Antila, MA, Project Manager, Museum Centre Vapriikki, kimmo.antila@tampere.fi
Harri Mattila, Principal Lecturer, HAMK University of Applied Sciences, harri.mattila@hamk.fi

Ch.3.
Vuokko O. Kurki, MSc (EnvEng), Doctoral candidate, TUT, vuokko.kurki@tut.fi
Osmo T. Seppälä, DTech, MD, Finnish Water Utilities Association, osmo.seppala@vvy.fi

Ch.4.
David Sauri, Professor of Human Geography, Department de Geografia, Universitat Autònoma de Barcelona, david.sauri@uab.cat
Leandro Del Moral, Professor of Human Geography, Departamento de Geografía Humana, Universidad de Sevilla, lmoral@us.es

Ch.5.
Troy Lepper, Ph.D. (Sustainable Development and Environmental Sociology), troylepper@yahoo.com

Ch.6.
Carol Fort, Senior Lecturer in History, Associate Dean School of International Studies, Flinders University, Australia, Carol.Fort@flinders.edu.au

Ch.7.
Johann Tempelhoff, Professor, School of Basic Sciences, North West University, South Africa, Johann.Tempelhoff@nwu.ac.za
Frank Winde, Professor of Geography, School of Basic Sciences, North-West University South Africa, frank.winde@nwu.ac.za

Ch.8.
Heikki S. Vuorinen, Adjunct Professor, Hjelt Institute, Department of Public Health, University of Helsinki, heikki.vuorinen@helsinki.fi

Ch.9.
Pekka E. Pietilä, DTech (CivEng), TUT, pekka.e.pietila@tut.fi

Ch.10.
Maria Rusca, Lecturer Urban Water Governance, Department of Integrated Water Systems and Governance, UNESCO-IHE Institute for Water Education, m.rusca@unesco-ihe.org.

Ch.11.
David Hall, director of the Public Services International Research Unit (PSIRU), Business School, University of Greenwich, London, D.J.Hall@gre.ac.uk
Emanuele Lobina, Principal Lecturer, PSIRU, London, E.Lobina@gre.ac.uk

Ch.12.
José Esteban Castro, Professor of Sociology, School of Geography, Politics and Sociology, Newcastle University, UK, esteban.castro@ncl.ac.uk

Ch.13.
Maarten Blokland, Associate Professor in Water Services Management, UNESCO-IHE Institute for Water Education, Delft, m.blokland@unesco-ihe.org

Ch.14.
Ezekiel Nyangeri Nyanchaga, Associate Prof., Department of Civil and Construction Engineering, University of Nairobi, Kenya, info@samezconsultants.com, enyangeri@uonbi.ac.ke

Ch.15.
David Nilsson, Dr., Affiliated Researcher, Department for Philosophy and the History of Technology, Royal Institute of Technology, david.nilsson@abe.kth.se

The thinking behind the book

Two photographs reflecting the thinking of the book: from official and formal arrangements of the pasts towards more participatory and democratic water governance and discussions. Above: Ground water investigations in Hämeenlinna in 1909 (photograph: Hämeenlinna Historical Museum) and Workshop in relation to the World Water Forum, Marseilles, France, March 2012 (photograph: Finnish Water Forum).

Preface

Various studies and policy documents clearly indicate that water resources and services will be among the major challenges to mankind in the future. If things continue as they are, close to half of mankind will face water quality or quantity problems by 2050. We must find alternative ways of improving the management and governance of current and future water systems and services.

While running and walking in different parts of the globe, the first editor has been able to quench his thirst from public taps and drinking fountains. Especially during long runs on hot days water gives the stamina to push further and reach your destination. Yet, one rarely thinks about the system needed to provide this vital element. Even less thought is given to water governance after a water break as the run continues. Only when running through the poorest parts of a city do such thoughts tend to enter one's mind when passing children and women carrying water in buckets to their homes. The rich and poor alike should have access to safe water. Water governance is like a marathon. Much preparation is needed before coming to the starting line. In water services the marathon starting line corresponds to a situation where safe water is available for everyone. That's where the race begins.

The process of running a marathon is very complex as is the operation of water utilities and services. An optimal result can be achieved in many ways, but no hand book solutions exist. The political, environmental, social, technical, etc. conditions should be taken into account while struggling towards the optimal goal whether we are running a marathon or dealing with water governance. It should also be noted that the journey does not end at the marathon finish line or when optimal water services are in operation. Constant maintenance is required to keep everything in good condition. There is no time for long pauses.

Since the 1980s, the second (co-ordinating) editor has increasingly dealt with water governance: management, institutional, policy and even political issues of water services. Technology and technical artefacts are certainly important, but a lot of other issues also need to be faced. One of them is the everlasting debate on water pricing and the broader question of how water services – both community water supply and sanitation, often also stormwater – should be managed in different conditions? Since political, economic, social, technological, environmental and legal conditions vary, the options for water services do also.

No universal solutions exist: each case has to be considered separately. Made decisions should be flexible enough for the operational environment that will change in the future while at the same time we should recognise and understand the limitations of earlier decisions about our systems. In spite of differences in the North and South, East and West, there seem to be also common things and issues – something that we can share and learn from each other.

During the years the third editor has worked in the water services sector, he has noted that most of the research conducted and literature written has been strongly focused on "solving" the problems of water services management. Given that we are still far from universal service coverage, the emphasis on providing such solutions is understandable. At the same time, many of the proposed solutions are quite ideological in nature. Success of solutions is often actively created and claimed with the aim of disseminating them. What lacks in this instrumental approach is critical reflection and research into how the sector is functioning. We agree that we must seek to develop solutions for the challenges facing the water services sector, but it is also important to conduct more critical empirical research on the functioning of the water services sector. Although there are a few notable exceptions among academics who engage in research which critically analyses the water services sector, they produce only a small share of the literature on the water services sector.

We, the editors, hope that this book and the 16 invited cases will for their part elaborate some key features of water governance and services in long-term perspectives. We are looking forward to reading further writings of the Governance and Management Series. It is also our hope that this book provides an impetus for further research critically analysing the institutions, policies and organisations which represent water services operating in different locations and conditions.

Foreword

Among the world's economic, social and environmental issues, the provision of water services and sustaining the water environment rank very high if not at the top. While some people regard water management as primarily a technological issue, history proves that institutions are as important as technologies and infrastructure in providing services across water supply, wastewater, and other water needs. Why this occurs is explained by the editors of this volume, which explains lessons about water services management and governance and now is welcomed to the IWA Publishing series on Governance and Management for Sustainable Water Systems.

Topics addressed in the series focus on institutions in the management of water, and they include important subjects such as governance processes, capacity-building, regulation and administrative law, social equity, and policy, among other topics. They reach across issues of sector governance such as water allocation and use; water supply and public health; wastewater and water quality; irrigation and drainage; instream flows; dam ownership; flood disaster management; and river basin and regional authorities. This volume focuses on water services, as the authors explain the nature of required governance by analyzing the shift in emphasis from technology to governance across a range of situations. These extend to water services across several continents and include discussions of management scenarios from urban water services to tradeoffs in water rights to environmental and health issues.

The reason that governance and management are so important is explained by the urgency of current global water issues that include water shortage and drought, rising sea levels, polluted water, floods, environmental degradation or other problems. As the first volume of this series explains, these problems point to the same conclusion: water must be managed effectively, fairly and sustainably to support survival and prosperity for all people and living things. This lesson has been taught successively in time, as the authors of this volume explain, and the lessons that resulted can inform us about how to manage and govern water sustainably in the future.

Ultimately, water management is a local issue and access to safe and reliable water services depends on local situations, even if they are influenced by global issues such as climate change. While many people have access to safe and reliable water services, billions lack access to it, and this leads to societal breakdowns, sickness, disorder, and environmental degradation. More than technology, this is the result of inadequate governance. As the World Water Assessment Programme explained in its 2006 report *Water: A Shared Responsibility*, in many countries water governance is in a state of confusion with a lack of water institutions or fragmentation of authorities and decision-making structures that result in many unsolved water problems that cause misery and poor living conditions.

The editors and authors are to be congratulated for adding this volume on lessons from history to this series. It is our hope that the knowledge and light provided by the analyses provided will advance the cause of Governance and Management for Sustainable Water Systems.

Neil S. Grigg
Colorado State University

Acknowledgements

The idea for this book originates from several meetings and discussions held with professional colleagues over the last decades. Once we learned about the new Book Series on Governance and Management by IWA Publishing, we immediately proposed this volume to address the issue of the relevance of history for any viable futures of water governance and management. We were very pleased to hear about the acceptance of this book for publication in the series.

The two editors have had the priviledge to conduct close to two dozen interdisciplinary studies on water utilities and their long-term development in Finland since 1998. Two EU supported projects, PRINWASS and WATERTIME, and activities related to the International Water History Association have also contributed to this book in several ways. Two research efforts in Delft, the Netherlands have also provided inputs for this book and underscore the need of using historical approaches to understand the current and future water services sector. On-going research (since 1997) on the Dutch water services sector and the UPaRF Water Supply and Sanitation Partnerships project (2008-2013) have also supplied various useful contributions to this book.

The voluntary contributions of all invited authors are gratefully acknowledged. We thank Maggie Smith and her colleagues at IWA Publishing for their kind advice and collaboration. Financial support from the Finnish Cultural Foundation (for this book and the Argumenta series 2011 – 12), Academy of Finland (Researcher mobility in working life: A Strategy That Works – Long-Term Lessons from Water Services in Finland, no. 135843; and AKVA Research Programme: Water as Social and Cultural Space: Changing Values and Representations, no. 263417), as well as Nordforsk Researcher Network 2011 (no. 40921 – Viable Water Management and Governance for Future) is highly valued. The support from the UNESCO-IHE Partnership Research Fund is also very much appreciated as is the indirect support from Fullbright New Century Scholars Program 2009/10 and the NSS funding from CIMO for the "Urban Water Services in Long-term Perspectives" project 2012-14.

The in-kind support from the home institutions of the editors, Tampere University of Technology and University of Tampere, Finland as well as UNESCO-IHE in Delft, the Netherlands is warmly acknowledged. Thanks are also due to many other colleagues and friends for their indirect contributions. We hope this effort will be useful to all of you and many others in the water sector.

We also express our thanks to Riikka Rajala who kindly acted as Assistant Editor and Sari Merontausta who drew several graphs.

We are looking forward to seeing further writings of the Governance and Management Series.

The Editors

Prologue

Petri S. Juuti, Tapio S. Katko and Klaas Schwartz

For much of the past century, the water services sector has been approached from a largely instrumental and technological perspective. Problems and challenges of expanding water services have largely been seen as technological problems. The technological approach has dominated also in developing countries. In the past decades, however, a more complex approach to the water services sector and the challenges it faces has gained ground. It considers different technological, socio-political and environmental dynamics and their interactions. As a result, the technological paradigm, though still prominent, is no longer hegemonic. Institutions are now considered as important as infrastructure. Moreover, the public sector, where water utilities frequently operate has not been static. The public sector, in general, has undergone considerable reforms in the past decades. In many countries the government has limited its role in providing water services, leaving a larger role for the private sector and civic society. In addition, society at large has also emphasised environmental issues.

The sixteen chapters of this book, divided in four sections, analyse the complexity of the water services sector based on a historical analysis of developments within the sector. The underlying conviction is that only by understanding past trends, processes and developments can the current situation in the water services be understood. And only through this understanding can policies for sustainable water services in the future be formulated. The four key sections relate to governance frameworks, technology and socio-ecological interactions, government and governance, and long terms policies. These sections are elaborated upon below.

GOVERNANCE FRAMEWORKS

When shifting from a technological approach to a broader governance approach, new frameworks for analysing that shift are required. In Chapter 2, such a framework, incorporating policies, actors and institutions, is presented and applied to the water services sector of Zambia. Chapter 3 presents an alternative framework, which examines changes in the water services sector by analysing historical changes using technology development theories. Chapter 4 examines the water services sector from the perspective of integration, and argues that provision of water supply and sanitation could in many cases be integrated, although less frequently so on the small or inter-municipal scale. What all these frameworks have in common is that they highlight shifts in the water services sector over the past decades. These shifts may be more governance-oriented, more technological in nature or related to the degree of integration of service provision. By highlighting these shifts, this section emphasises the dynamic nature of the water services sector and the interaction of the socio-political, technological and environmental dimensions.

TECHNOLOGY AND SOCIO-ECOLOGICAL INTERACTIONS

The management of water resources and the provision of water services are tightly interwoven into the social and environmental context in which they take place. The infrastructure used for managing resources and providing services impacts society and the environment, which, in turn, impact the infrastructure and actual management and provision of water. In the chapters of section two this interwovenness of the infrastructure for management of resources and the provision of services is elaborated upon. First, the governance of large infrastructure in Spain is examined by David Sauri and Leandro del Moral. In this chapter different modes of governance are linked to different eras in water infrastructure development. Secondly, the complicated trade-offs in management of water resources in the state of Colorado are investigated by Troy Lepper. The third chapter discusses the sustainability of water resources use in South Australia. The fourth chapter of this section looks at the problem of acid mine drainage in South Africa, where toxic pollutants from worked-out mine operations pose a major threat to water resources. The section ends with an elaboration on the usage and safety of lead pipes for the provision of drinking water. This elaboration incorporates both a historical perspective on the use of lead pipes as well as an assessment of current practices.

FROM GOVERNMENT TO GOVERNANCE

An important element of the technological paradigm is the prominence of the single public service provider, which provides standardised and universal services. In this approach, little room exists for non-state actors in the provision of services. With the shift from government to governance, however, private organisations and civic society have become increasingly prominent in the service provision process, even if the fundamental overall public responsibility for ensuring provision of water services can hardly be questioned.

In this section the interaction of government organisations and non-state entities in the provision of water services is brought to the forefront. First, the various roles and responsibilities of local municipal governments in different European countries are discussed by Pekka Pietilä. This overview highlights the strong diversity in the roles played by local government in the provision of water services. The second chapter of this section presents a historical perspective on partnerships in the provision of water services. It discusses different partnership forms, such as twinning arrangement, public-private partnerships and water operator partnerships, and places these arrangements in a particular prevailing social and political context. David Hall and Emmanuel Lobina examine the rise and decline of multinational water companies in the next chapter. In the subsequent chapter Esteban Castro discusses the role of citizenship within the current governance of water services. He particularly analyses the obstacles and possibilities for the "substantive democratisation of the management of water and sanitation services".

LONG-TERM POLICY

The final section of the book deals with long-term policies in the water services sector and analyses related shifts – or the lack thereof – over time. The first chapter looks at the centralisation and internationalisation of the Dutch water supply sector. The authors argue that the driving forces behind the implemented policies have shifted from government-led policies to policies pronounced and pursued by the management of water utilities. Although these policies have over time led to increased centralisation of the water supply sector, the reasons for pursuing more centralised provision are quite different. In the second chapter water services management and governance in Kenya is analysed by Ezekiel Nyangeri Nyanchaga. The aim of this historical analysis is to identify lessons for improving current and future service provision. David Nilsson undertakes a historical analysis of donor policies in the water services sector. He comes to the conclusion that few donors have been able and willing to learn from previous experiences in the water services sector. Finally, the section contains a thorough analysis of the reform of water services provisioning in the Helsinki Metropolitan Area. The discussion of these reforms reveals that political drivers appear more relevant in explaining the nature of the reforms than expert views of water professionals.

All in all, water governance is an exercise in political, economic, administrative and social authority, which influences the development and management of water resources and related services delivery as pointed out by the 1st edition of the United Nations World Water Development Report in 2003. It comprises mechanisms, processes and institutions through which citizens and groups articulate their interests, exercise their legal rights, meet their obligations, and mediate their differences in relation to water resources. – We argue, however, that cross-sectional and historical intra- and international comparisons are a valuable method of study of different sectors of human life, including technologies and governance.

Section I

Water Governance Frameworks

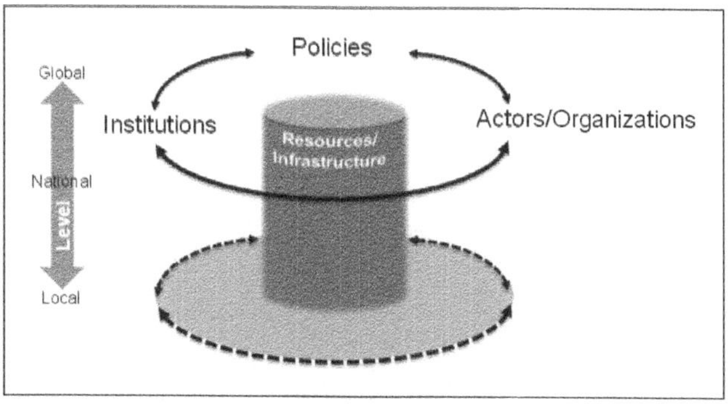

Section I, Water Governance Frameworks consists of three chapters. The 1st chapter of this section was written by three authors, Kemerink J., Mbuvi D. & Schwartz K. It focusses on governance shifts in the water services sector through a case study of the Zambian water sector.

The 2nd chapter was written by three authors, Antila K., Katko T. S. & Mattila H. This theoretical paper concentrates on water services evolution and technology development theories.

The 3rd chapter was written by five authors, Katko T. S., Kurki V. O., Juuti P. S., Rajala R. P. & Seppälä O. T. It investigates the integration of water and wastewater utilities through a case study Finland.

Chapter 1

Governance shifts in the water services sector: a case study of the Zambia water services sector

Jeltsje Kemerink, Dorcas Mbuvi and Klaas Schwartz

1.1 INTRODUCTION

The public sector has been increasingly subject to reforms over the past decades. The pace and scope of change to the structures and dynamics of the state, in general, have become much greater than previously experienced (Thynne, 2000; See also Pollitt, 2003). The main thrust of these reforms is a shift from the hierarchical state and direct government management through the application of laws and regulations towards "greater reliance on horizontal, hybrid and associational forms of government". That implies a more inclusive government which involves a broader spectrum of actors including the private sector, non-profit organizations, or other government agencies (Hill & Lynn, 2005:172). Swyngedouw (2005:1992) has referred to this shift as a move to "governance-beyond-the-state", which concerns the "emergence, proliferation and active encouragement [...] of institutional arrangements of "governing", which give a much greater role in policy-making, administration and implementation to private economic actors, on the one hand, and to parts of civil society, on the other, in self-managing what until recently was provided or organized by the national or local state".

Underlying this shift from government to governance was the advent of neo-liberalism in the 1970s as a reaction to the "overloaded State" (Skelcher, 2000). In the neoliberal perspective the hierarchical mode of governance was condemned as inherently inefficient and was to be replaced by an alternative mode of governance, preferably a market mode (Bevir & Rhodes, 2001). In market governance, the market essentially functions as an allocative mechanism through which decisions regarding the management and allocation of resources are taken. The underlying idea is that markets "empower citizens in the same way as we exercise power as consumers" (Pierre & Peters, 2000:19).

The water services sector has been no exception to this general trend. This trend is visible in a variety of private sector institutions[1] (such as markets, efficiency, performance management and, principles of quasi-competition) which dominate management practices in the current water services sector. In this

[1]In this paper we use the term institutions in a sociological sense meaning "as a rule, norm, or custom simultaneously enabling and constraining human agency" (Bakker, 2002).

chapter we examine this trend and how these governance shifts have manifested themselves in the water services sector. For this purpose we present a framework for analyzing governance shifts and apply this framework to the case of Zambia. We find that in the case of Zambia, although the momentum of involving private sector organizations in the provision of water services may have diminished, private sector institutions have increasingly been incorporated in "public" service provision modalities.

1.2 A FRAMEWORK FOR ANALYSIS

Shifts in the governance of water services are analyzed by using a framework of four interdependent elements: institutions, organizations, policies and the actual infrastructure (or resources). One characteristic of this framework is that it adheres to the distinction between organizations and institutions (Bakker, 2002). Organizations in the water services sector are the actors which are assigned responsibilities for providing services, developing policies and/or regulating the sector, or any other actor which influences the decision-making processes of the sector. Institutions are understood in the sociological sense of the term. Institutions are social arrangements that shape, regulate and reproduce human behavior across time and space. These social arrangements may be formal in nature but can also be informal. At any given time and location multiple institutions may co-exist and possibly conflict with each other. A third element of the framework concerns water policies. These policies may be explicit (for example, the National Water Policy of 1994) but may also be more implicit or largely symbolic in nature.

Organizations, institutions and policies stretch across multiple scales. In discussing the nature of "water politics"[2] Mollinga (2008) identified various domains in which social power relations converge, mediate and shape water management in and across states. The domains[3] consist of a global domain, a domain of the politics of water policy, and a domain of everyday politics. The global domain refers to water discourse, policy and tentative regulation at the international level. Actors in this political arena are organizations which are internationally active in the water services sector. They include, among others, development banks, donor countries, UN agencies, international interest groups, international NGOs and academics. They form a "network of water policy elites" (Conca, 2006) that discusses, and at times advocates, policy prescriptions for reforms in the water services sector. The policy domain concerns the policy processes regarding water services in the context of the sovereign state. Water policies are "negotiated and re-negotiated in all phases, stages and at all levels, and are often transformed on their way from formulation to implementation" (Mollinga, 2008:12). National water organizations (ministries, utilities, national NGOs, etc.) are the main actors in this domain. The "everyday" domain, as we use it in this chapter, concerns the daily service provision of water services at the local level.

The framework emphasizes the continuous interaction between the various elements of the governance framework and is to a large extent dynamic and subject to minor and major shifts. Major shifts will be reflected as changes in the institutions, organizations, policies and possible infrastructure and technology choices. Minor shifts will be noticed as changes in one of the elements but may not lead to changes in other elements. The elements at various scalar levels interact strongly, and a governance structure can be analyzed in terms of the prevailing mix of institutions, organizations, policies and technology that predominate in the water services sector at a given moment in a specific location (Figure 1.1).

[2]Mollinga distinguishes the different domains to discuss water politics. Our purposes of citing these domains (identifying different scalar levels) are somewhat different. However, the distinction between the domains is useful in our view.

[3]Mollinga (2008) developed this framework to analyze the politics of water resources management where he also distinguished the domain of inter-state hydropolitics. In the context of urban water supply, however, this domain seems less relevant.

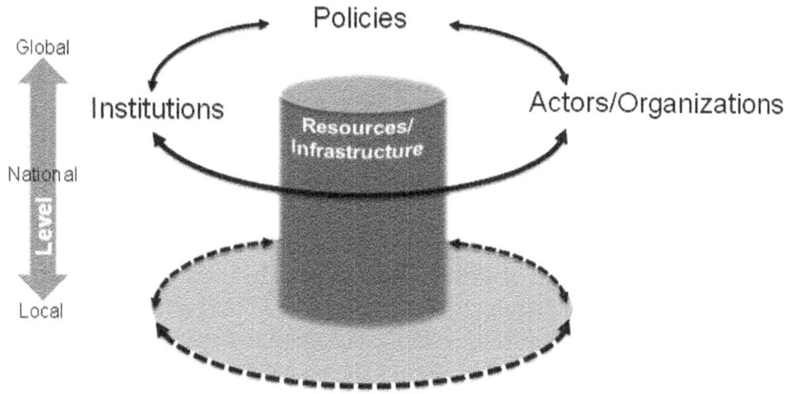

Figure 1.1 Framework for analyzing governance shifts.
Source: Kemerink and Schwartz, 2010.

1.3 GOVERNANCE SHIFTS IN THE WATER SERVICES SECTOR

Despite the ambitious target of providing safe water and sanitation for all, the Drinking Water Decade (1981–1990) ended in disappointment. Considerable investments were made in infrastructure, but at the end of the Decade as many people lacked access to safe drinking water and sanitation as when the Decade began (Economist, 1994). The main lesson learnt from the decade was that the provision of water supply and sanitation was not just a technological and capacity problem. Increasingly, importance was given to "institutions".

With the onset of neo-liberalism in the 1980s, this emphasis on institutions was quickly translated to mean private sector involvement. The start of the concession contract in Buenos Aires in 1993 marked the unofficial start of the "water privatization decade" (Franceys, 2008:45). Between 1993 and 2003 private sector participation was heavily promoted by a coalition of international development banks, bi-lateral donors and a group of water sector professionals. Advocacy for private sector involvement was based on a number of arguments. First of all, private sector involvement was believed to increase the efficiency of a utility by making it operate under the profit motive and by binding it to a contract containing clear performance objectives. Secondly, private sector involvement separates regulation from actual service provision, thus addressing the poacher-gamekeeper[4] problem. Thirdly, private operators were expected to directly finance the investments required to improve services, thus relieving the government from subsidies to the water supply and sanitation sector. Fourthly, experience from privatization in England and Wales "played a major role in convincing policy makers that private financing of urban water utilities could be viable" (Marin, 2009:19; Franceys, 2008).

Since the turn of the century, however, the initial optimism of those supporting increasing involvement of the private sector in developing countries became tempered. High-profile concession contracts once presented as representing the future of the water services sector collapsed or faced increasing operational difficulties. Terminated concession contracts include the El Alto-La Paz concession (2005) in Bolivia and the Buenos Aires concession (2005) in Argentina. Other contracts, such as the concession contracts in Manila and Jakarta faced serious operational challenges in the past decade

[4]The poacher-gamekeeper problem concerns the concentration of regulatory and operational functions within the same public sector entity. The poacher-gamekeeper problem leads to weak regulation and accountability of the service provider.

(Jensen, 2005; Braadbaart, 2007). Illustrative of the declining popularity of private sector involvement in developing countries is the 2003–2004 Action Plan of SUEZ, one of the world's largest private water operators, where the company emphasizes that it will "concentrate on the soundest markets providing the most recurrent revenues starting with the Franco-Belgium domestic market and including the European Union and North America". The Action Plan further explains that "exposure to emerging countries, as measured by capital employed, is expected to be reduced by close to one third" (SUEZ, 2003).

A number of arguments have been forwarded to explain the difficulties faced in establishing functioning public-private partnerships (PPP) in the water services sector. Hall and Lobina (2005) argue that high transaction costs undermined the successfulness of PPP projects. These transaction costs include the legal, consulting and financial costs of establishing a PPP, managing the risks involved in the PPP and the establishment of regulatory agencies responsible for regulating the private operator. Hall and Lobina argue that such costs can amount to 10% of the total costs of a project. Braadbaart (2005) argues that the design of the "first generation" PPP contracts in the 1990s was conceptually flawed[5]. In his assessment these flaws actually led to a situation where PPP projects were not followed through, failed tenders and suspended contract negotiations. Those contracts that were awarded had to be renegotiated shortly after they became effective. Other authors have noted the lack of adequate regulatory capacity as a severe shortcoming of many PPP projects (Parker & Kirkpatrick, 2005; Prasad, 2006). Some have noted opportunistic behavior of the private operators characterized by "profit-seeking and risk avoiding behavior of international water companies, in interaction with local and national governments" (Lobina & Hall, 2003:3–4).

The end of the water privatization decade did not mean that the water services sector returned to the service provision modality which characterized the decades before the onset of private sector involvement. Although service provision through public utilities became accepted again, and the World Bank even decided to "re-engage" with public service providers (Schwartz, 2006), public utilities increasingly were expected to operate according to private sector management practices and principles. On the other hand, concepts such as "private institutions" (Bakker, 2002), "market organized production" (Swyngedouw, 2004) and "private sector ethos" (Smith, 2004) have become increasingly important in the water services sector.

1.4 GOVERNANCE SHIFTS IN THE ZAMBIAN WATER SERVICES SECTOR

In this section we examine the governance shifts which took place in the Zambian water services sector in the past decades. Firstly, the policy reforms introduced in the 1990s are elaborated upon. The policies incorporated principles which strongly influenced the organizational and institutional changes in the sector. The third section looks at changes in institutions guiding the sector. As reflected in the policy principles, the sector shifted increasingly to more private sector institutions. The third part of this section looks at the establishment of Commercial Utilities and an independent regulator as the organizational form through which commercialization of the Zambian water sector has taken place.

1.4.1 Policy reforms

Soon after the International Drinking Water Supply and Sanitation Decade, the first multi-party government was elected in Zambia. Providing water services remained a challenge to the new government. The Lusaka

[5]See Braadbaart (2005) for the five "flaws" of the first generation PPP contracts.

Province, for example, received less than 2 hours of daily water supply (Nyumbu *et al.* 1997). More than 60 percent of the distributed water was lost through illegal connections. About 72 percent of the then formally connected customers in Lusaka Province were discontent with the service delivery reliability, sufficiency and quality (Nyumbu *et al.* 1997).

With the support of, respectively, GTZ (German Technical Cooperation) and KfW (German Development Bank), Lusaka and Chipata local authorities transformed their water schemes into water and sewerage companies in 1989 and 1992. These commercial utilities were purposely established as pilots in an experiment to test whether and how local authorities could cost-effectively provide universal and affordable quality water supply services and increase revenues. However, the water board members elected to the newly established companies were mainly political appointees such as the city mayor or town clerks. These political appointees were hardly familiar with the operation and management of a water utility. As a result, it was hard to hold the board members accountable, routine maintenance was neglected, and eventually the water systems deteriorated. Sector subsidization was likened to "*putting money into a hole*"[6]. At the same time, the economic downturn which had been besetting the country since the 1970s[7] meant that the Zambian government had few funds to invest in the water services sector. Between 1990 and 1995 the budget allocation for investment in the water service sector dropped from US$ 4.3 million to US$ 0.9 million (Nyumbu *et al.* 1997).

Additional reforms were initiated shortly after the Lusaka and Chipta utilities were established. Consensus existed between the various stakeholders in the Zambian water sector that the causes of poor performance could be traced back to weaknesses in the country's legislative, institutional and organizational setup (Chanda, 2000). Reforms to address these weaknesses were initiated in 1993. These water sector reforms were linked to other reform and privatization campaigns which swept through Zambia under the auspices of the World Bank (Cocq, 2005) strongly reflecting the private sector participation focus forwarded by the World Bank. By 1994 the government of Zambia had adopted the National Water Policy and the Strategy and Institutional Framework for Water Supply and Sanitation (developed by a Programme Coordination Unit (PCU) established by the Cabinet), which spelled out the principles according to which the water supply and sanitation sector would be organized. The policy principles put forth in these documents were as follows:

- Separation of water resources management from water supply and sanitation
- Separation of regulatory and executive functions within the water supply and sanitation sector
- Devolution of authority to local authority and private enterprises
- Full cost recovery – in the long run
- Human resource development leading to more effective institutions
- Technology appropriate to local conditions, and
- Increased government priority and budgetary allocation to the sector.

With the principles of the reform strategy laid down, the PCU established the Water Sector Development Group (WSDG) in 1994, which was tasked with implementing the reforms as outlined by the PCU in the sector principles (WUP, 2000). The WSDG prepared legislation for the water supply and sanitation sector with the aim of providing a legal framework for the reforms. A first draft was prepared by 1995, and following a period of consultation, a revised draft was submitted to the Ministry of Legal Affairs three months later. After another period of 18 months, during which the proposed legislation was

[6]Personal interview.

[7]Per capita income fell from US$ 752 in 1965 to US$ 351 in 2002 (Dagdeviren, 2008:103).

discussed by the Ministry of Local Government and the Ministry of Energy and Water Development, the Water Supply and Sanitation Act was passed by parliament in 1997 and signed by the President in November 1997 (WUP, 2000).

1.4.2 Institutions

Based on the policy principles set forth in the National Water Policy, private sector institutions, such as emphasizing markets, efficiency gains, performance management and principles of quasi-competition, were introduced in the water services sector. Achieving cost-recovery has been a prime focus of the reforms implemented in Zambia. Dagdeviren (2008:104) even argues that cost recovery is the "core objective" of commercialization in Zambia. The original target of the Zambian Regulator of water providers, the National Water Supply and Sanitation Council (NWASCO), was to achieve full cost recovery by 2010 (NWASCO, 2005). Cost recovery was to be achieved through increased user charges and efficiency. In 2005, NWASCO approved tariff adjustments for all but two utilities. The adjustments were 20% to 50% increases[8]. Since NWASCO became operational in 2000, tariffs have increased by 200% to 400% (NWASCO, 2005).

Competition is also promoted in the Zambian water sector. Its main objective is to stimulate efficiency gains in the sector. Competition is introduced through a benchmarking exercise facilitated by NWASCO. The main results of this "competition" are made public in the annual comparative sector report produced by NWASCO. The idea behind the benchmarking exercise and the comparative sector report is that it "induces competition by benchmarking water utilities which leads to increased efficiency and enhanced performance. Through this benchmarking exercise, each utility "is motivated to improve its own previous performance as well as to outperform the other CUs" (NWASCO, 2005).

The policies introduced in the 1990s have also highlighted the need of allowing utilities to operate at arm's length of the government. Given Zambia's inability to sustain subsidization of the water services sector, "agencification" of service providers was pursued. Agencification refers to "the conversion of government-departments which previously operated in a hierarchical chain [...] into semi-autonomous [agencies]" (van Donge, 2002:315). These utilities, operating at arm's length of the government, are expected to reap the efficiency gains associated with private sector organizations[9]. NWASCO, for example, argues that organizing utilities as autonomous agencies "gives more promise to the achievement of cost recovery leading to more sustainable provision of services" (NWASCO, 2006:6).

1.4.3 Organizations of the Zambian water services sector

The policies implemented in the 1990s had a considerable impact on the organizations involved in various dimensions of service provision in the Zambian water services sector. A key element of the National Water Policy was that it called for the separation of regulatory functions from those of actual service provision. The regulatory functions included tasks such as licensing service providers, defining the service area and standards for the service providers, regulating the tariffs that utilities could charge, monitoring service protection and imposing sanctions for failure to meet set standards. Tasks attributed to the service providers were development of infrastructure for service provision, operating and maintaining that infrastructure and overall management of provision of water services. In Zambia, it

[8]The average tariff charged by the commercial utilities fluctuates between US$ 0.21 and US$ 0.25 depending on the quantity of water consumed (NWASCO, 2005).

[9]The question whether empirical support exists for the view that private sector organizations are more efficient than public organizations is beyond the scope of this research. This view, based on the property rights theory, is, however, often forwarded (Braadbaart, 2002).

was decided that separating the regulatory tasks from service provision tasks would be done by introducing the so-called "British model" of regulation (Foster, 1996). The 1997 Water Supply and Sanitation Act established an independent regulatory agency, The National Water and Sanitation Council (NWASCO), which became operational in 2000. The main tasks of NWASCO are to license water utilities, establish and enforce guidelines for the development and management of water utilities, advise utilities and other service providers and disseminate information to consumers (GTZ, 2006). In this capacity NWASCO regulates all water utilities, private or public. NWASCO can enforce its functions through the licensing and tariff setting process. The regulator is largely financed by the service providers themselves, who are to contribute 2% of their turnover in the form of a license fee to financing NWASCO. In 2003 (when the contribution by the licensees was 1% of their turnover), the license fee made up 72% of NWASCO's budget. The remaining 18% was financed by the government and donors (NWASCO, 2004a).

The actual service providers also changed as part of the reforms. The new organizational framework devolved water supply and sanitation responsibilities to Local Authorities. The act provided the Local Authorities a number of options for arranging service delivery in their locality. The Local Authorities could:

- Provide services through a Unit or Department within the Local Authority.
- Establish a Commercial Utility (CU). In this scenario a local authority would establish a government-owned company (either on its own or jointly with other local authorities) and transfer responsibilities for providing services to this government-owned company.
- Involve the private sector in providing services[10]. The local authority could involve the private sector in providing services by a range of PSP contracts (including setting up a joint venture with a private company).

The option that eventually was selected was that of establishing commercial utilities, which although government-owned, would operate as private companies under Zambian Law and according to commercial principles. The preference for commercial utilities is strongly linked to the adherence to private sector institutions described in the previous section and in particular the target of achieving full cost-recovery.

Table 1.1 highlights two developments. First of all, the commercial utility has become the dominant organizational mode for providing services in the sector. Service provision by central or local government has been faded out. Secondly, the utilities providing services at the moment are much larger in scale than the ones that provided services prior to the reforms. The total number of service providers has decreased from 84 to 17 organizations. The reduction in the number of service providers is mainly due to the fact that commercial utilities spanning multiple local authorities have been established in the intervening period. Richards *et al.* (2008) highlight the importance of commercial utilities as a tool to pursue commercialization, but specifically note the need for these utilities to operate at a certain scale. By letting commercial utilities service multiple municipalities the provider can "enable the realization of synergies and economies of scale and thus improve prospects for commercial viability" (Richards *et al.* 2008:20). As such, the organizational setup of the sector is strongly linked to the policies and institutions guiding the sector (Table 1.2).

[10]In the late 1990s, a study was undertaken (financed by the World Bank) concerning the possibility of involving the private sector in service provision in the capital of Lusaka. This study (undertaken by the English private operator Severn Trent) found that involving the private sector by way of a lease contract for the city of Lusaka was considered the most suitable reform strategy. The study was subject to heated discussion, and eventually was not acted upon (Cocq, 2005).

Table 1.1 Water providers in Zambia in 1997, 2004 and 2009.

	Before 1997	2004	2009
Central government	46	0	0
Local authorities	29	23	0
Commercial Utilities	2	51	11
Private schemes[a]	7	10	6

[a]The majority of private water suppliers in Zambia are companies (such as Zambia Sugar PLC, Chilanga Cement and Zambia Revenue Authority) which supply water to their employees as a fringe benefit.
Sources: NWASCO, 2004; NWASCO, 2010.

Table 1.2 Shifts in institutions, organizations and policies in the Zambian water sector.

Governance structure	Established as at 1993	Established by 2010
Institutions	– Non-autonomous providers. – Non-existent performance management & benchmarking practices. – No (full) cost recovery. – Self-regulation.	– Autonomous agencies (2000). – Performance management through SLG & A (2001). – Benchmarking by use of partial indicators (2002). – Quasi-competition (2002). – Incentive regulation (2008). – (Full) cost-recovery (2008).
Organizations	– Water boards as surface water abstraction regulators (1946). – DWA under MWLNR as water service developers & suppliers across district townships. – LAs as water service providers across municipal/city councils. – Other multi-providers as water service providers especially, where DWA and LAs services were lacking. – Public Service Review & Reorganization Commission (1993).	– Inter-ministerial PCU (1995). – Water sector development group (1995). – NWASCO as an independent regulator (2000). – MEWD & MLGH mandates separated (2000). – Public commercial utilities (2000). – Water watch groups (2002). – Devolution trust fund (2003). – Part-time inspectors (2005).
Policies	– Local Government Act Cap 281 (1991).	– National Water Policy (1994, revised in 2010). – Water Supply and Sanitation Act (1997).

DWA: Department of water affairs; MWLNR: Ministry of water, land and natural resources (that transformed into the MEWD in 1992); LAs: Local authorities; Multi providers: Other ministries such as the ministry of local government and housing (MLGH), private companies, local and international organizations; LWSC: Lusaka water & sewerage company; CWSC: Chipata water & sewerage company; SLG & A: Service level guarantees & agreements; RIP: Regulation by incentive program; PCU: Programme coordination unit (that later transformed into NWASCO); MEWD: Ministry of energy and water development.

1.5 CONCLUSIONS

The reforms implemented in the Zambian water supply sector in the past two decades mirror those introduced in other countries in Sub-Saharan countries such as Uganda, Namibia, Kenya and Mozambique. Under pressure from international financing agencies and donors, policies to introduce private sector participation were developed in the 1990s. Although the policies often called for private organizations to provide services, actual privatization was postponed and eventually suspended altogether as too little support at the national and local level existed for it. Allowing for privatization in the policy and legal framework guiding the sector is one thing, but actually replacing public water utilities with private operators appeared to be a bridge too far. At the same time, the incorporation of private sector institutions in the sector continued. In Zambia, the sector is currently characterized by efforts to achieve full cost–recovery, quasi-competition, performance management and regulation of commercial utilities through an independent regulator. The inclusion of private institutions is strongly promoted and supported by international donors. At the same time, public employees of the water utilities and representatives of relevant government organizations have maintained a preference for keeping the water utility within the public domain. Incorporating private sector institutions is one thing, but outsourcing of management and operation of the water utility to the private sector is another.

REFERENCES

Bakker K. (2002). From state to market? Water mercantilizaciốn in Spain. *Environment and Planning A*, **34**(5), 767–790.

Bevir M. and Rodes R. (2001). A Decentered Theory of Governance, Rational Choice, Institutionalism and Interpretation. Institute of Governmental Studies University of California, Berkeley, http://www.escholarship.org/uc/item/0bw2p1gp?display=all#page-2. (Retrieved February 8, 2010).

Braadbaart O. (2002). Private versus public provision of water services: does ownership matter for utility efficiency? *Journal of Water Supply: Research and Technology-AQUA*, **51**(7), 375–381.

Braadbaart O. (2005). Privatizing water and wastewater in developing countries. *Water Policy*, **7**(4), 329–344.

Braadbaart O. (2007). Privatizing water: the Jakarta concession and the limits of contract. In: A World of Water: Rain, Rivers and Seas in Southeast Asian Histories, P. Boomgaard (ed.), KITLV Press, Leiden.

Chanda O. (2000). Regulating the Water and Sanitation Sector in Zambia. National Water Supply and Sanitation Council, Lusaka.

Cocq K. (2005). “There is still no alternative”: the beginnings of water privatization in Lusaka. In: The Age of Commodity: Water Privatization in Southern Africa, G. Ruiters and D. McDonald (eds.), Earthscan Canada, Toronto.

Conca K. (2006). Governing Water: Contentious Transnational Politics and Global Institution Building. MIT Press, Cambridge.

Dagdeviren H. (2008). The commercialization of urban water services in Zambia. *Development and Change*, **39**(1), 101–121.

van Donge J. (2002). Agencification. In: Handbook on Development Policy and Management, C. Kirkpatrick, R. Clarke and C. Polidano (eds.), Edward Elgar Publishing, Northampton.

Economist (1994). Of wets and water. *The Economist (London)*, March 26th.

Foster V. (1996). Policy Issues for the Water and Sanitation Sectors. IDB, Washington, D.C.

Franceys R. (2008). GATS, ‘privatization’ and institutional development for urban water provision. *Progress in Development Studies*, **8**(1), 45–58.

Hall D. and Lobina E. (2005). The Relative Efficiency of Public and Private Sector Water. Public Services International Research Unit, www.psiru.org (accessed 28 November 2011).

Hill C. and Lynn L. (2005). Is hierarchical governance in decline? Evidence from empirical research. *Journal of Public Administration Research and Theory*, **15**(2), 173–195.

Jensen O. (2005). Troubled partnerships: Problems on coping strategies in Jakarta's water concessions. Paper presented at the 4th Conference on Applied Infrastructure Research in Berlin on 8/10/2005. http://www.infraday.tu-berlin.de/fileadmin/documents/infraday/2005/papers/jensen_Troubled_Partnerships_Problems.pdf. (Retrieved 19 August, 2008).

Kemerink J. and Schwartz K. (2010). Shifts in Governance. Unpublished Lecture note. UNESCO–IHE Institute for Water Education, Delft.

Lobina E. and Hall D. (2003). Problems with Private Water Concessions: A Review of Experience. Public Services International Research Unit, www.psiru.org (accessed 18 December 2011).

Marin P. (2009). Public-Private Partnerships for Urban Water Utilities. World Bank, Washington, D.C.

Mollinga P. (2008). Water, politics and development: framing a political sociology of water resources management. *Water Alternatives*, **1**(1), 7–23.

NWASCO (2004). Water Sector Reform in Zambia. NWASCO, Lusaka.

NWASCO (2005). Urban and Peri-Urban Water Supply and Sanitation Sector Report 2004/2005. NWASCO, Lusaka.

NWASCO (2006). Urban and Peri-Urban Water Supply and Sanitation Sector Report 2005/2006. NWASCO, Lusaka.

NWASCO (2010). Urban and Peri-Urban Water Supply and Sanitation Sector Report 2009/2010. NWASCO, Lusaka.

Nyumbu I., Mphande C., Pertz K., Cummins K., Helming S., Mndege N., Mutasa E., Jareg P. and Ravdal E. (1997). Review of the Water Sector Reforms Programme in Zambia: Main Report. Government of the Republic of Zambia: Programme Co–ordination Unit.

Parker D. and Kirkpatrick C. (2005). Privatisation in developing countries: a review of the evidence and the policy lessons. *The Journal of Development Studies*, **41**(4), 513–541.

Pierre J. and Peters B. (2000). Governance, Politics and the State. Palgrave Macmillan, Hampshire.

Pollitt C. (2003). The Essential Public Manager. Open University Press, Maidenhead.

Prasad N. (2006). Privatisation results: private sector participation in water services after 15 Years. *Development Policy Review*, **24**(6), 669–692.

Richards T., Doering E., D'Souza A., Lang H., Plumm H., Schaefer D. and Werchota R. (2008). Water Supply and Sanitation Sector Reforms in Kenya, Tanzania, Uganda and Zambia: Challenges and Lessons. German Development Cooperation, Frankfurt.

Skelcher C. (2000). Changing images of the state: overloaded, hollowed-out, congested. *Public Policy and Administration*, **15**, 3–19.

Smith L. (2004). The murky waters of the second wave of neoliberalism: corporatization as a service delivery model in Cape Town. *Geoforum*, **35**(3), 375–393.

SUEZ (2003). SUEZ Introduces its 2003–2004 Action Plan: Refocus, Reduce Debt, Increase Profitability. Press Release. SUEZ, Paris.

Swyngedouw E. (2004). Social Power and the Urbanization of Water: Flows of Power. Oxford University Press, Oxford.

Swyngedouw E. (2005). Governance innovation and the citizen: the janus face of governance-beyond-the-state. *Urban Studies*, **42**(11), 1991–2006.

Thynne I. (2000). The state and governance: issues and challenges in perspective. *International Review of Administrative Sciences*, **66**, 227–240.

WUP (2000). LWSC's Institutional Arrangements for Dealing with Peri-urban Areas. WUP, Abidjan.

Chapter 2

Technology development theories and water services evolution

Kimmo Antila, Tapio S. Katko and Harri Mattila

2.1 INTRODUCTION

Water and sanitation services are basic infrastructure services in all communities. They have been for long the most important water use in urban areas, although demand has been created by various factors over time. In the case of Finland, a Nordic country, demand was created particularly by the need for fire fighting and drinking water and health and hygiene requirements.

Historically the development of water supply and sanitation, or water services management, has gone through several phases. Early on, in the middle of the nineteenth century, it was referred to as public health engineering, the emphasis being on public health requirements. Later in the century terms such as municipal engineering, sanitary engineering and public works engineering emerged. Environmental engineering has been introduced more recently (Barraqué *et al.* 2008). Sometimes the partial business nature of water services has been reflected in the use of the term "water industry".

In many countries public water supplies were developed first for urban areas followed by sewerage systems. However, in Finland both we built more or less at the same time. The first systems were established in the biggest cities and townships, and they spread gradually to smaller townships as the population of the country grew (Katko, 1997). Contrary to urban areas, diffusion of water supply in rural areas started with small systems (Katko, 1997), and by 1940 some 70 rural systems had been established. The development was linked to the order in which various technologies spread to farmhouses: electricity came first, followed by cowshed water supply, dwelling sewer, and finally dwelling water supply (Katko, 1992).

In the Finnish context, management of water has four distinct levels (Figure 2.1). The highest level, in terms of geography, consists of various types of inter-municipal systems, most of them operating on wholesale basis in both water supply and sewerage. There were some forty such entities in 2010 (Pietilä *et al.* 2010a). The next level includes those serving townships and cities and surrounding population centres.

For long water supply and sanitation services were managed by separate organisations, but since the 1970s and 1980s many of them, particularly in urban areas, have merged into a single entity, at least in Finland and Sweden (Katko *et al.* 2010). According to a comparative study on the evolution of water and sewerage services in 29 European cities in 13 countries, mergers of water and sewerage utilities are quite uncommon in the European context. For instance, in Germany water and gas utilities have traditionally worked in close co-operation or been merged, whereas wastewater services have been part of city technical services (Stadtwerke). (Juuti & Katko, 2005) From the hydrological cycle point of view, it is obvious that merging of water and wastewater systems could be the first natural step towards integrated water resources management. Integrated urban water management (IUWM) has been promoted as a good system for example, by the EU-supported SWITCH project (van der Steen & Howe, 2009) and Mitchell (2004).

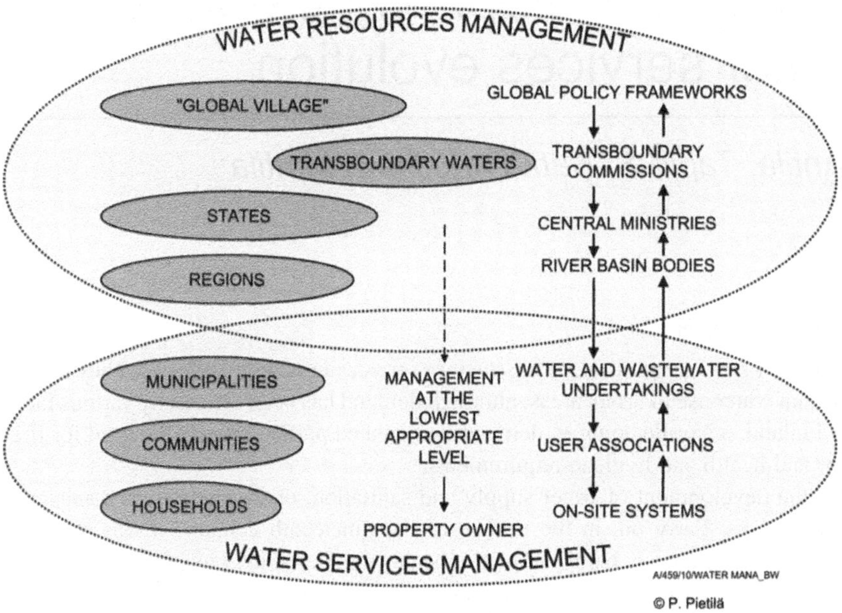

Figure 2.1 Management of water resources and services at various levels in Finland.
Source: Pietilä, 2006.

The third level consists of small rural systems commonly managed by co-operatives which are gradually also getting more involved in wastewater services (Takala *et al.* 2010). The lowest level includes on-site water and wastewater systems typically serving one or a few households outside of networks. This level is important in a country that still has a fairly large number of people living in dispersed rural areas. Finland also has some half a million holiday homes which have to have proper sanitation by 2016.

International water management research and discussion have hardly touched on the linkages between water resources management and water and wastewater services. The concept of total water management introduced by Grigg (2008) deals with this dilemma and the necessity to examine these interrelations. It is likely that climate change will raise integrated water resources management to an even more important role than it has today.

Water supply and sanitation have several special characteristics compared to many other commodities and services. In addition to the technical and economic dimensions, they have to take into account also social, cultural, administrative and legal traditions and practices that are often of vital importance. Water supply and sanitation are a natural monopoly in most cases, as originally suggested by John Stuart Mill in 1848 (Sharkey, 1982). Under a monopoly, it is sensible to construct only one system for a service area. In some cases water has to be transported from long distances as in the coastal areas in Finland. Generally we have more or less local or regional resources available.

Water supply and sanitation are unique in the sense that they have many stakeholders each with their preferences, priorities and professional capabilities. Conflicts may rise between the various interest groups or even within them. It is also possible that the interests of central government differ from those of regional and local authorities. The current European Union water policy also has its own implications. The mentioned four different levels of water and waste water systems also increase the diversity. Yet, despite the obvious complexity, a multiple stakeholder paradigm (Hukka & Katko, 2009) of this kind certainly promotes flexibility. Another special feature of water services is its long-term nature – they are expected to operate still a hundred years from now (Melosi, 2000). Water and sewage systems are highly capital-intensive, while water is an economic as well as a public good with a mixture of other features that vary by time and place. (Pietilä *et al.* 2010b). As Merkel *et al.* (2011) point out, water supply is strongly influenced by local, structural conditions – natural and man-made ones that any water utility has to cope with in its operational area.

It is now time to explain the aim of this paper after the introduction of water services: to discuss the relevance of various technology development theories and how such theories could be applied particularly in the field of water supply and sanitation. The examples of water management are mainly from Finland but there are also findings from developed and developing economies.

2.2 WHAT IS TECHNOLOGY?

It is understandable that the focus in technology and classical engineering studies is on technological artefacts and principles. Yet, Leppälä (1998) defines technology more widely to cover (i) *technological artefacts,* (ii) *procedures, and* (iii) *knowledge* required to apply (i) and (ii). This wider approach to technology is highly appropriate as regards water and water services, as will be discussed below.

Futures researchers carrying out studies on a variety of fields and issues often use PESTEL (political, economic, social, technological, ecological and legislative) analysis. Its advantage is that it makes researchers assess the development of technology or other sectors in a wider context. It often seems that we focus on technical and economic issues only when trying to assess technology.

In the late 1970s, before the United Nations International Drinking Water Supply and Sanitation Decade (IDWSSD) 1981–1990, Pacey (1977) discussed the dimensions of *appropriate technology*, and concluded that technology alone is not enough, but that we also need a variety of criteria for technical, social and economic appropriateness. Pacey (1983) also introduced two major spheres for examining technologies: the *user sphere* and the *expert sphere* arguing that "good technology" should make use of both of these major spheres (Figure 2.2). The development of technology is biased if only the views of the expert sphere are taken into account. Thus, it is important to consider also the user sphere and take into consideration cultural and organisational aspects.

As a practical example, Pacey (1983) mentioned the maintenance of hand pumps in the conditions of developing countries. During the IDWSSD 1981–1990, a world-wide development project on appropriate hand pumps that could be operated as close to the village level as possible was carried out. It involved manufacturers, external support agencies, laboratory tests and field experiments (UNDP, 1984). In that context a mixture of high and low technology was used. The village-level operation and maintenance (VLOM) hand pump has to be as simple as possible from the viewpoint of the users, whereas from the manufacturer's perspective it should be high technology since it is advisable to use the best possible materials. One of the best hand pumps developed then was the

NIRA direct action pump that could be maintained at the village level. Often older ladies proved to be the best caretakers.

After the introduction of water supply and sanitation and the examination of technology, we shall now deal with the various technology development theories and discuss how they can be used to explain the evolution of water supply and sanitation.

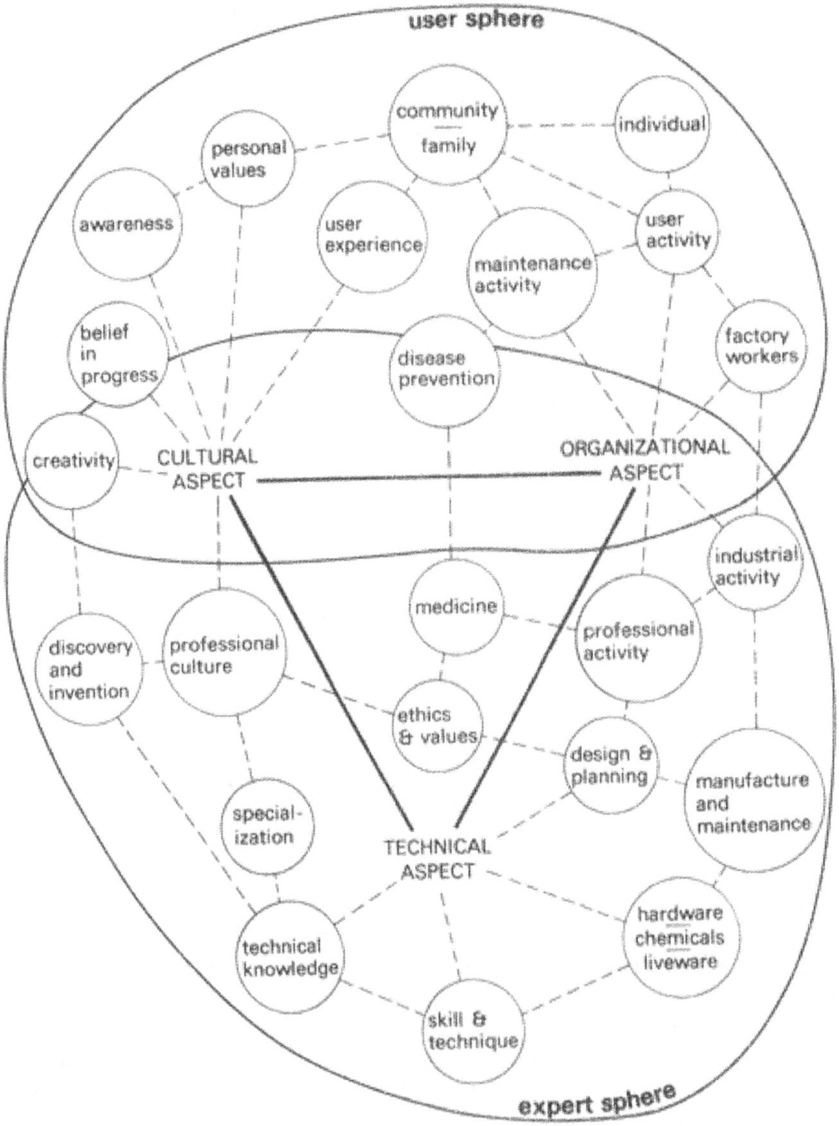

Figure 2.2 User and expert spheres and their cultural, organisational and technical aspects regarding technology development.
Source: Pacey, 1983, reprinted with the permission of John Wiley & Sons.

2.3 TECHNOLOGY DEVELOPMENT THEORIES VERSUS WATER MANAGEMENT

2.3.1 Institutional economic theory

Before moving to technology development theories, we will have a look at *institutional economic theory*. Hodgson (1988, Figure 2.3) points out how for example, economic co-ordination is never merely a price issue, but it should be supported by a wide range of other economic and social institutions. The institutional approach, therefore, involves both technology and individual tastes and preferences. The concept of "socio-economic system" emphasises that the economy is inseparable from social and political institutions and is thus largely applicable to water management.

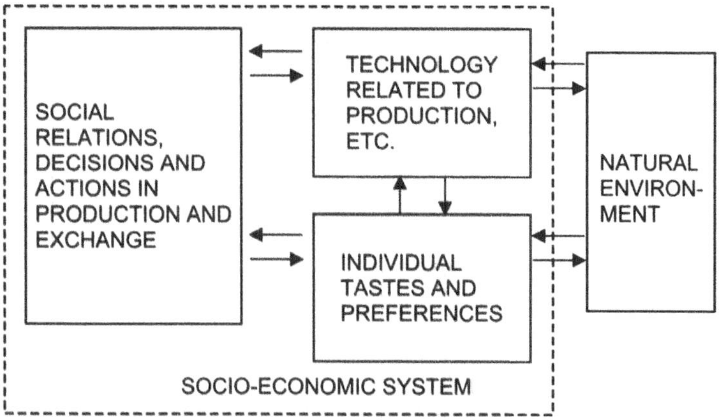

Figure 2.3 Projected domain of institutional economic theory by Hodgson (1988, reprinted from *Economics and Institutions,* with the permission of Polity Press).

The 1993 Nobel Laureate Douglas C. North (1990, 3–4) defined New Institutional Economics (NIE) and institutions as flows: "Institutions are the rules of the game in a society, or the humanly devised constraints that shape human interaction" while "organisations are groups of individuals bound to some common purpose to achieve objectives". Using the analogy of soccer, institutions are the rules of the game while organisations are the players (North, 1990:3). It is interesting to note that several Nobel Laureates in economics have embraced the institutional approach including Douglass C. North (1993), Joseph E. Stiglitz (2001), Paul Krugman (2008), and Elinor Ostrom (2009).

It is important to look backwards and think about the technological past which could give deeper insights into the question of technology. History of technology is not merely study of inventions and innovation processes but much more. David Edgerton also emphasises an important fact about water and sanitation systems: it is not clear that a new technology is always superior to the an old one. It is also noteworthy that some old technologies can co-exist with new ones or even replace them. (Edgerton, 2007).

The evolution of water and sanitation services and related technological changes can be assessed by a variety of technology development theories. In this paper we will explore the following approaches: technological determinism; Kranzberg's laws; large technological systems (LTS); reverse salient; path-dependence theory; technological momentum; evolutionary development; technology jumps; social construction of technology (SCOT); and others. In many respects they all are linked to sustainability.

2.3.2 Technological determinism and Kranzberg's Laws

Technological determinism was for long a major stream of thinking. It was based on the idea that technology will develop continuously more or less in the same direction in the absence of disturbances or changes in the development paths. In spite of the fact that history of technology experts proved decades ago that technological determinism cannot explain the real nature of technology development, it still seems to be a common way of thinking – if not mainstream – in technology policy and decision-making. For instance, in Finland most policy makers and even several experts still seem to think that water and sewerage systems can expand without limitations. Such deterministic thinking is, however, being gradually replaced by more critical views as will be discussed later.

In 1985 Melvin Kranzberg presented the so-called *Kranzberg's Laws* (Table 2.1). Firstly, he argued that technology as such is neither good nor bad, nor is it neutral, but is rather a combination of them depending on the conditions. The first law of Kranzberg states that although technology as such might be a fundamental element in many instances, non-technical factors and motivations are always more important in technology policy-related decision-making. Thus, technology is very human by nature as is the history of technology.

Table 2.1 Kranzberg's Laws.

Law I	Technology is neither good nor bad; nor is it neutral
Law II	Invention is the mother of necessity
Law III	Technology comes in packages, big and small
Law IV	Although technology might be a prime element in many public issues, nontechnical factors take precedence in technology-policy decisions
Law V	All history is relevant, but the history of technology is the most relevant
Law VI	Technology is a very human activity, and so is the history of technology

Source: Kranzberg, 1986.

History of technology is also a story of man and tools – hand and mind – working together. Technological innovations and methods typically have various environmental, social and human consequences that depend heavily on the principles of technical equipment and systems. Besides, technology can also produce different impacts in different conditions and contexts of use.

By his first law, Kranzberg wanted to remind us that it is important to compare the consequences of short-term and long-term thinking. We should compare "what could have happened" with "what actually happened". That is possible only if we realise that the impact of technology and collaboration varies depending on values, institutions and cultural environments. The second Kranzberg law argues that each technical innovation requires other technical systems to become fully operational and as efficient as possible. The emergence of technology in packages, big and small, is pointed out by the third law. Accordingly, modern technological systems normally consists of a combination of several types of processes and components.

The fourth law of Kranzberg is probably more complex than the others – even a dilemma. Engineers have traditionally argued that proposals are based on technological facts and not on sentimental or otherwise qualitative interpretations. The first mentioned facts were based on input–output thinking. Although political issues play a bigger role in decision-making compared to mere technological issues, there is no cause for concern since that is the essence of a democratic society. To cope with this interaction between technology and environment, we have developed various social tools to assess technologies.

The question arises whether these new developments should be better taken into account in engineering education, and even in student selection processes which currently emphasise mathematical and physics skills. Or should we just leave these matters for continuing education and life-long learning?

As a history of technology expert Kranzberg argued that all history is relevant while the history of technology is most important. His fifth argument is interesting since, at least from the engineering point of view, it looks like it has been largely ignored. Kranzberg claims that one of the reasons why school children and students often show little interest for history is that the mainstream of history research does not value or recognise the technological dimension. Yet, in real life, all of us can see the consequences of technology.

The sixth Kranzberg law states that technology is very human by nature as is the history of technology. All technical artefacts are used by an engineer, a worker, a businessman, a soldier, etc. The fundamental purpose of technology is human as is its misuse. Kranzberg reminds us that, for instance, in computer engineering hardware is totally useless without software.

2.3.3 Large technical systems

The theory on *large technical systems* (LTS) developed by Thomas P. Hughes (1987) deals with systems or networks of enormous proportions or complexity such as electrical systems. According to Hughes, large technical systems evolve through five major phases. The first phase of an LTS involves an invention like those of Thomas Alva Edison. In the second phase the technology has to be developed into a marketable product (systems builder phase). The third phase of technology transfer includes the transfer of the system to various locations and societies. It also forces applying the system to new areas with their requirements described by the term "technological styles". The fourth "systems growth" phase was beset by the problem of direct current (DC) in the case of Edison. The solution was to introduce alternating current (AC) that forced changing the technological system and also led to new social innovations. (Hughes, 1991).

Hughes also uses the concept of *"reverse salient"* that can technical or cultural in nature. The term has a military origin, where it refers to a backward bulge in an advancing front (Figure 2.4). In the fifth phase of an LTS the system has acquired "dynamic energy" and is capable of resisting external pressures.

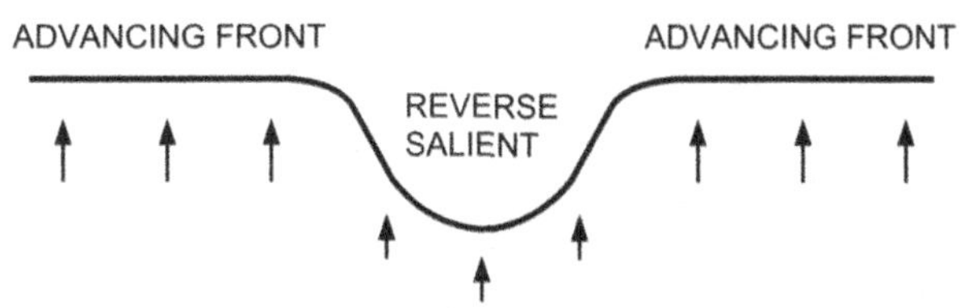

Figure 2.4 Reverse salient. (*Source*: Hughes, 1991, reprinted with the permission of Nat Acad of Eng. Publ. Nat Acad Press.)

The LTS theory is widely applicable to electrical systems and grids that still seem to be growing bigger and becoming interconnected, for instance, in the European context. The theory is also widely applicable to road systems and networks – national and transnational ones. However, in water supply and sanitation the applicability of the LTS theory is quite limited since water supply and sanitation are to a large extent based on utilising local water resources although there are some long-distance water transfer systems dating back all the way to Roman times (Mehlhorn & Weiss, 2009). The Spanish Hydrological Plan presented in the early 1980s involved such a system where the aim was to construct 60 to 70 dams along the Pyrenees and transfer water to the southern parts of the country. It aroused resistance and led to the establishment of the New Water Culture Foundation and the signing of the European Declaration for A New Water Culture in Madrid in 2005.[1] Thus, the LTS theory is applicable to water management only to some extent.

2.3.4 Evolutionary development

Futures researchers, or futurists as they were called earlier, have pointed out the *evolutionary nature of development.* Accordingly, development and technology are not deterministic and bifurcations or turning points are bound to occur. In terms of water, bifurcation also refers to upstream water bodies of watercourses from which waters may flow to two different catchment areas. Figure 2.5 illustrates that principle. Sometimes we may also encounter *revolving paths* like in the selection of urban raw water sources in Finland. The first systems used mainly ground water, but around 1920 many switched to surface water since the ground exploration and drilling technology available then was considered inadequate. After WW II there was a new interest in using surface water, but use of ground water was gradually reintroduced accompanied by artificial recharge since the 1970s. Now we seem to be stuck with the groundwater paradigm although surface waters have improved dramatically due to efficient water pollution control. In any case, such a change and return toan "original" path can easily take decades or -longer.

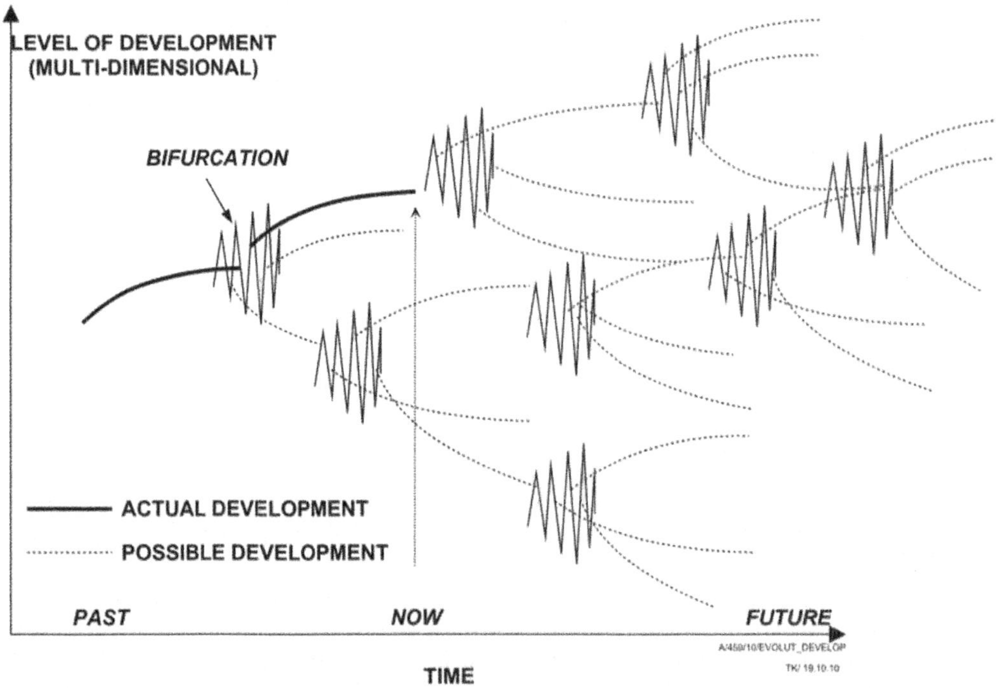

Figure 2.5 Bifurcations in technological development, sometimes involving revolving paths. *Source*: Mannermaa, 1991, modified by Katko, 2012.

[1]European Declaration for a New Water Culture; http://www.unizar.es/fnca/euwater/index2.php?x=1&idioma=en

In literature, path dependence (Ch. 1) is often explained through technological log-ins: in many cases it is presented in a negative light, as something that has limited our options or led to a too fixed path. Yet, such decisions can be either negative or positive, depending on whether the decisions prove to be clever in the long run or not. One positive example of a positive decision – not to use lead pipes in Finland – is presented in Chapter 5. Now, some 130 years later, many of the cities in the western world still have huge problems with lead pipes and water quality since pipe replacement requires time and resources.

2.3.5 Visionary management

The basic idea of futures researchers is to identify a *vision* – a stage of development that is ideal and feasible. The aim is to develop various alternative scenarios and strategies for reaching those scenarios. As Bell (2003) points out, current futures research is based on active development instead of the earlier futuristic thinking which was based more on modelling the futures. Hence, the work of researchers and decision-making will have a major influence on the kinds of futures we will have. Therefore, it is important to identify possible, credible and preferable futures and development paths (Seppälä, 2004). It is also worthwhile analysing the analogies between "prehistory of futures" and the "pasts of presents".

2.3.6 Technological momentum and jump

The fifth Kranzberg law introduced the concept of *technological momentum*. Hämeenlinna, a city in southern Finland, decided around 2000 to establish a large supra-municipal water and wastewater company to provide services to several municipalities. The decision was taken in a situation where the neighbouring municipalities were willing to collaborate with the central city of Hämeenlinna. The municipalities had earlier experiences of collaboration in other sectors. The discussion about wide-spread municipal mergers in Finland, which grew in intensity during the decade, had also begun. One important applied tactic was was to offer the smaller neighbouring municipalities the majority of the seats on the board leaving the biggest city in the minority. This allayed the concerns about the bigger city dictating decisions: technological momentum was expanded into *political momentum*. Later, due to changes brought about by municipal mergers, that merger of the water utilities would have been much more difficult. Thus, political momentum is largely dependent on making decisions at the right time.

Technology can develop very fast through so-called *technology jumps*. Early on, rural water supply in Finland was typically based on wooden drilled logs for water distribution. The logs were locally available and cheap but, particularly in winter time, due to temperature changes, high enough water pressure could not be maintained without many leakages. Continuous repairs were also troublesome. Then the manufacture of plastic pipes started in Finland in the mid-1950s – only small sizes were made initially. After experience had been gained, it was also possible to produce larger plastic pipes. Nowadays polyethylene and PVC pipes dominate in Finnish water supply and sewerage systems and their relative share in new water supply and sewerage systems has been probably the highest in the world for long.

Another technology jump was the introduction of metering. It was taken into use in Germany and most of the Western countries in the nineteenth century, whereas England started to introduce metering more than 100 years later, after the privatisation of the systems in 1989. Although England was the forerunner of water supply and sanitation in the nineteenth century, in metering they certainly were latecomers.

Yet, the question of who decides water tariffs and levels and on what grounds seems to remain unanswered. Local politics largely determine tariffs and metering. Rehabilitation of a water utility is often of secondary importance for decision makers. From the ethical viewpoint it is right that each generation bears its reasonable share of responsibility.

An example of political influence is the development of the legislation concerning various types of pollution discharged into water courses. Over the decades, increasingly smaller pollution sources have been moved from the category of diffuse pollution to point source pollution and thereby out of control. Earlier small amounts of wastewater were considered diffuse pollution, but today they are controlled more (Mattila, 2005). The development of legislation has led to a situation where one might ask whether the concept of diffuse pollution is still relevant in Finland (Mattila, 2009).

2.3.7 Social construction of technology

The central elements of the *social construction of technology* (SCOT) theory were initially developed by Bijker, Hughes and Pinch (1987) and developed further by Bijker (1995) and McKenzie and Wajcman (1999). The last-mentioned team wonders why "technological change is often seen as something that follows its own logic" – something we may welcome, or protest against, but are unable to alter fundamentally.

The main focus of the SCOT theory is trying to understand the links between social and technical processes and seeing them as a human or social construction. According to the theory, technology is shaped by human engineers, market forces, consumer needs and demands as well as all individuals and groups who are also social products. Other important issues are power politics and the internal structure of technology. The SCOT theory contends that we have different solutions to different problems that are related to the needs and concerns of specific groups.

Mattila (2005) used the SCOT approach to analyse small on-site sanitation systems in rural Finland. He showed that acceptable solutions can only be arrived at through negotiations between different social groups. A basically simple issue such as rural sanitation may get complicated leading to some other solution than the very best one being finally selected for a site. The first choice is between water-based and non-water based sanitation. In Finland it is self-evident that due to the low population density piped water supply and sewerage cannot be extended everywhere. Therefore, we will have a considerable number of permanent residences and recreational homes relying on on-site systems also in the future.

As Mattila (2005) points out, in addition to developing technological systems we should try to do our best to come up with proper management systems. In modern society it is unlikely that people themselves ever would like to run their own toilet or wastewater systems. Therefore, we should try to promote alternative systems for proper operation and maintenance. Figure 2.6 describes the "soccer analogy" of Douglas C. North, where each player should assume a proper role and follow the rules of the game. It also shows the large number of on-site sanitation "players"; and there are even substitutes (actors) on the bench. Centralised wastewater management systems probably have fewer players.

It is interesting to note once again how the development of water supply and sanitation was politicised in Finland in 2010. The parliamentary elections in the spring of 2011 inspired politicians to speak and act to change (mainly add exemptions to) the Government Decree on Treating Domestic Wastewater in Areas Outside Sewer Networks at a time when implementation of the original decree had just begun: persons who were not sanitation experts were writing on the subject with great conviction.

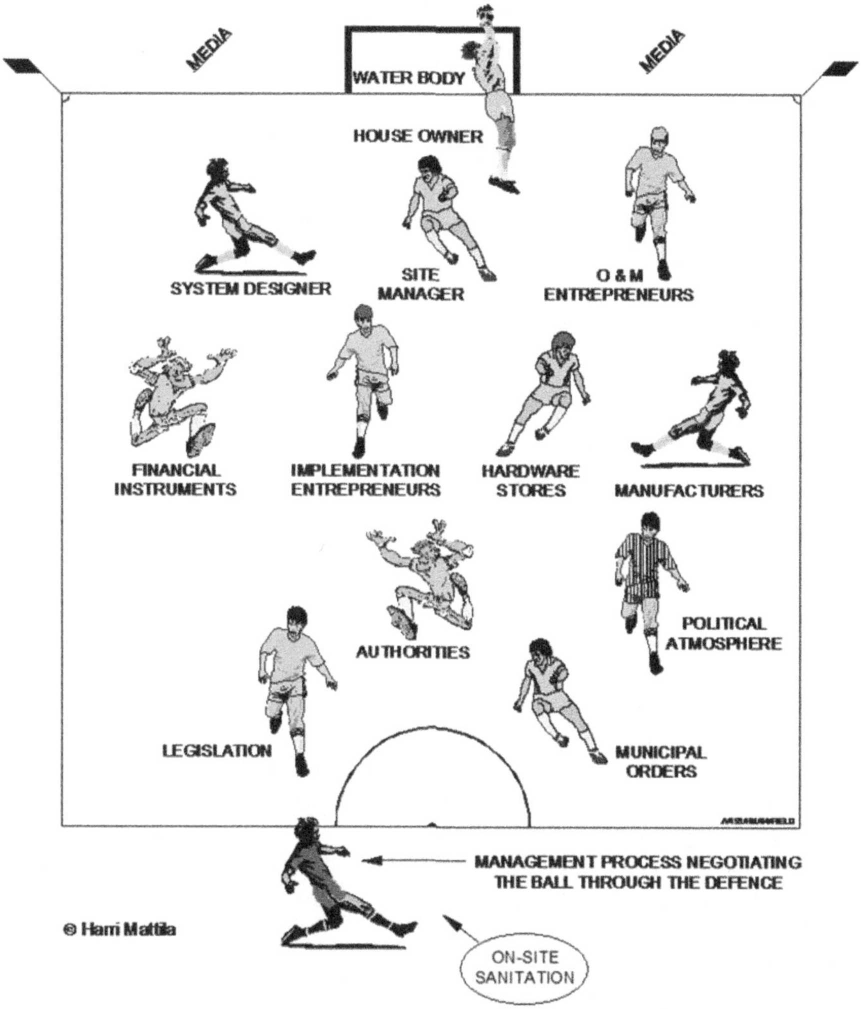

Figure 2.6 Various players involved in on-site sanitation: the other side can be stopped by team work between the players (stakeholders); in the absence of proper team work the game will become paralysed. *Source*: Mattila, 2005:83.

2.3.8 Issues of governance

The international *privatisation* of water supply and sanitation promoted strongly by international financial bodies and multinational companies in the 1990s was a paradox in terms of history and history research. The promotion happened to a large extent in an unhistorical context – without recognising the largely negative experiences that almost all countries had in the late 1800s. In fact, by 1900 a great majority of the systems had been taken over by local governments in most Western countries. Yet, the private sector has always played an important role as supplier of services such as consultancy, construction and other operational services. (Hukka & Katko, 2003).

The various roles of the private and public sector and other related major parties obviously may change over time and by location. In this context, it is probably advisable to look at various types of *stakeholder theories*. At the end of the day, long-term public interest will be the determining factor for sustainable water and sanitation systems and their management. The fact that the views of municipal residents as the final beneficiaries should be taken more seriously is also being understood better.

The concept of the *invisible city* refers to anything that normally easily goes unnoticed in daily city life. In terms of technology, invisible refers to that part of infrastructure that is non-visible, in many cases situated underground such as the bulk of water supply and sewerage systems. Kramer (1998) mentions that most developed countries have 50- to 100-year-old underground utility systems that carry far greater loads than designed for and are difficult to repair or replace. He suggests trenchless technologies for rehabilitating these systems. However, it is obviously a far greater challenge to make the invisible city more visible in decision-making and to the public.

Pinch (2010) refers to invisible technologies and points out that the sociology of technology has much to offer to historians. He notes that technology is so all-pervasive in our everyday world that we scarcely notice it. An important part of it is technical infrastructure: sanitary systems, power grids, roads and more recently the Internet. He echoed David Edgerton's call to move away from innovation-driven studies to studying *mandate technologies* of everyday life and opening up the black box of the mundane world by exploring the studies by Ervin Goffman. At least the second and third authors can approach the subject by asking their students what do they normally do first thing in the morning after waking up. You guessed right, most probably go to bathroom.

The various approaches above described the path to the diversity *of* the *technology–society complex* and relationships. Rip and Kemp (1998) explored the technology–society complex. They argue that "technological regimes, the coherent complex of scientific knowledge, engineering practices, production process technologies, product characteristics, skills and procedures, and institution and infrastructure in terms of a certain technology or mode of work organisations" are the central issues. They further regard technological regimes a broader, socially imbedded version of technological paradigms. They recognise three levels of systems: from micro to meso and macro levels. This framework by Rip and Kemp can potentially be used to analyse the evolution of water services that are typically managed and provided at various levels, from on-site to small community-wide and supra-municipal systems of various kinds and their paradigm changes.

Gabriel Hecht has used the concept of *technopolitics* to refer to the strategic practice of designing or using technology to constitute, embody, or enact political goals. Systems could in these cases also be hybrids of technology and politics or "politically constructed technologies". Decision made for these systems could not be reduced to just technologies or just political practises. (Hecht, 2001) Technopolitical regimes consist of linked sets of people, engineering practices, technological artefacts, political programmes, and institutional ideologies which act together to govern technological development and technopolitics. Regime refers to the people who govern and their ideas and hidden or more visible values are essential to future decision making. Hecht has analysed French nuclear power plants based on these methodological aspects and emphasised the importance of cultural forms in decision making.

More recently *vulnerability and risks* have been rising concerns in relation to technical systems and particularly water management. They are partly related to climate change but also to the aging of infrastructure systems. We have to remember that most of the systems were established some 150 years ago. In the case of Finland, the majority of the current networks were constructed in the 1960s and '70s during urbanisation and rural migration. We also have unfortunate warning examples of recent cases where piped water became contaminated in larger urban areas, although the problems are quite minor compared to the water-related diseases that many of the citizens in developing economies still have to struggle with.

2.4 DISCUSSION

As shown above there are different types of technology history research approaches and options. We have presented a selection of such theories, part of which could be used for analysing and understanding the evolution and governance of water services. Naturally, there are also other options and approaches not covered by this paper.

One of the key issues concerning history of technology research and its relevance for today's development and for futures assessment is related to its timeframe. The impact of technology and the importance of certain episodes or decisions depend largely on our own thinking in terms of time. In an operative timeframe it is possible that even relatively small changes or episodes seem dramatic whereas if we use a strategic or visionary timeframe for our futures and pasts, we may note that there may have been some other episodes that may be relatively much more important.

The study of history of technology is important because it provides an opportunity to accumulate lessons learned and apply them, although they are not necessarily always utilised to the extent that they could. In principle, we should continuously assess the impacts and consequences of our actions. Through continuous "check-ups" we should examine our experiences related to our current strategy and action plans. They could then be made the basis of our future aims and policy objectives.

Water management on global scale consists of various levels. The global village level, the third level, involves certain global policy frameworks or principles. The interstate or regional level deals with transboundary waters, both surface and ground water. The state level is where the central government operates and the regional level administers riverbasin bodies. The upper levels concern mainly water resources management, whereas water services management typically takes place at a lower hierarchical level. Some water supply systems are inter-regional like in the USA, while other supra-municipal, municipal, small community, or household level systems. As we adhere to the principle of proper policy frameworks and agreements, we also follow the principle of managing the services at the lowest appropriate level.

2.5 CONCLUDING REMARKS

Our examples of the various theories on history of technology and of water and sanitation reveal some more general governance principles that we find valid and important:

(i) Plurality: we have various options, only seldom just one
(ii) Diversity: we have several institutional options concerning various political, economic, social, technological, ecological and legislative dimensions
(iii) Locality: particularly in water supply and sanitation, the local actors and conditions are most important for sustainable systems and stakeholders
(iv) Globality: there is clear need for more general global principles that we can agree on and implement
(v) Politicality: many of the decisions related to water policy and management are political by nature, which should not be forgotten when exploring history of technology and its development
(vi) Environmentality: the environment is a key requirement for sustainability, but we must remember that it is not our property, but on loan from future generations

(vii) Humanity: technology as shown by the examples is largely human by nature
(viii) Visionarity: taking history into account, we should not base our decisions on short-sighted thinking but consider the longer term.

Water supply and sanitation systems seem to be more locally bound than several other infrastructure systems. Therefore, many of the theories of technology development are only partially applicable to the assessment of the evolution of water and sanitation services. It is also possible that some of the theories applicable to other fields of technology are not at all feasible for water services.

It is obvious that the timeframe and long-term development – the pasts and the futures – must be understood and considered more when setting technology policy. The same applies to engineering education and related policy making. Water management operates in an environment that is becoming ever more complicated and at the same time more vulnerable and increasingly unpredictable. In all of these activities, better understanding of the pasts and futures is of fundamental importance.

Acknowledgements

Advice by Osmo Seppälä and the financial support from Academy of Finland (Decision No. 135843) are gratefully acknowledged.

REFERENCES

Barraqué B., Formiga Johnsson R. M. and Nogueira de Paiva Britto A. L. (2008). The development of water services and their interaction with water resources in European and Brazilian cities. *Hydrol. Earth Syst. Sci.*, **12**, 1153–1164.

Bell W. (2003 Paperback). Foundations of Futures Studies I: History, Purposes, Knowledge. Transa. Publ., New Brunswick.

Bijker W., Hughes T. P. and Pinch T. (eds.) (1987). The Social Construction of Technological Systems: New Directions in the Sociology and History of Technology. MIT Press, Cambridge, MA.

Bijker W. (1995). Of bicycles, bakelites and bulbs: Toward a Theory of Sociotechnical Change. London MIT Press, Cambridge, MA.

Edgerton D. (2007). The Shock of the Old Technology and Global History Since 1900. Oxford Univ. Press, New York.

Grigg N. S. (2008). *Total Water Management: Practices for a Sustainable Future.* AWWA Publ., Denver.

Hecht G. (2001). Technology, politics and national identity in France. In: Technologies of Power, B. Michael, T. Allen and G. Hecht (eds.), MIT Press, Cambridge/London.

Hodgson G. M. (1988). Economics and Institutions. A Manifesto for a Modern Institutional Economies. Polity Press, Cambridge and Philadelphia.

Hughes T. P. (1987). The evolution of large technological systems. In: The Social Construction of Technological Systems: New Directions in the Sociology and History of Technology. W. Bijker, T. P. Hughes and T. Pinch (eds.), MIT Press, Cambridge, MA.

Hughes T. P. (1991). From deterministic dynamos to seamless-web systems. In: Engineering as a Social Enterprise, H. E. Sladovich (ed.), Nat Acad of Eng. Publ. Nat Acad Press, Washington, DC.

Hukka J. and Katko T. (2003). Water Privatisation Revisited – Panacea or Pancake? IRC, Delft. www.irc.nl/page/6003 (accessed 15 June 2012).

Hukka J. and Katko T. (2009). Complementary paradigms of water and sanitation services: lessons from the Finnish experience. In: Water and Sanitation Services: Public Policy and Management, E. Castro and L. Heller (eds.), Eartscan, London, UK; Sterling, USA, pp. 153–172.

Juuti P. and Katko T. (eds.) (2005) Water, Time and European Cities. History Matters for the Futures. WaterTime, EU.

Katko T. (1992). The development of water supply associations in Finland and its significance for developing countries. Discussion paper no. 8, The World Bank, Water Supply and Sanitation Division.

Katko T. (1997). Water! — Evolution of Water Supply and Sanitation in Finland from the mid-1800s to 2000. FIWA, Helsinki.

Katko T. S., Kurki V. O., Juuti P. S., Rajala R. P. and Seppälä O. T. (2010). Integration of water and wastewater utilities: a case from Finland. *J. AWWA.*, **102**(9), 62–70.

Kramer S. R. (1998). The invisible infrastructure. *Civ. Eng.—ASCE*, **68**(10), 48–51.

Kranzberg M. (1986). Technology and history: "Kranzberg's Laws". *Technol. Cult.*, **27**(3), 544–560.

Leppälä K. (1998). Miten tekniikkaa oikein tieteellisesti tutkitaan? (How to study technology in scientific ways?). *Tiedepolitiikka.*, **15**(2), 25–30 (In Finnish).

Mannermaa M. (1991). Evolutionaarinen tulevaisuudentutkimus (Evolutionary Futures Research). *Acta Futura Fennica no. 2* (In Finnish). APK-Kustannus, Helsinki, 362 p.

Mattila H. (2005). Appropriate Management of On-site Sanitation. Doct. diss. no. 537., Tampere University of Technology, Tampere.

MacKenzie D. and Wajcman J. (eds.) (1999) The Social Shaping of Technology. 2nd ed., Open Univ. Press, Buckingham, UK.

Mehlhorn H. and Weiss M. (2009). Long-distance water supply and linked systems in water supply. *GWF*. 150 (Special Anniversary Issue. W), 74–94 (In German).

Melosi M. V. (2000). The Sanitary City. Urban Infrastructure in America from Colonial Times to the Present. Johns Hopkins Univ. Press, Baltimore; Maryland.

Merkel W., Staben N., Bräcker J., Levai P., Neskovic M., Petry D. and Weiss M. (2011). Structural conditions in German water supply. *Water Sci. Technol. – Water Supply*, **115**, 596–604.

Mitchell V. G. (2004). Integrated Urban Water Management. A Review of Australian Practise. CRISO and AWA Report CMIT-2004-075.

North D. C. (1990). Institutions, Institutional Change and Economic Performance. Cambridge Univ. Press, Cambridge.

Pacey A. (1977). Technology is not enough: the provision and maintenance of appropriate water supplies. *Aqua*, **1**(1), 1–58.

Pacey A. (1983). The Culture of Technology. Blackwell, London.

Pietilä P. (2006). Role of Municipalities in Water Services. Doct. diss. no. 617., Tampere University of Technology, Tampere.

Pietilä P., Katko T. and Kurki V. (2010a). Vesihuolto kuntayhteistyön voiteluaineena (Water fueling municipal collaboration). Foundation for Municipal Development. Publ no. 62. (In Finnish).

Pietilä P., Katko T. and Seppälä O. (2010b). Uniqueness of water services. *E-Water*. http://www.ewaonline.de/portale/ewa/ewa.nsf/home?readform&objectid=7BBE56D63E38CEBDC1257727002F494A (accessed 15 June 2012).

Pinch T. (2010). The invisible technologies of Goffman's sociology. *Technol. Cult.*, **41**, 409–424.

Rip A. and Kemp R. (1998). Technological change. In: Human Choice and Climate Change, S. Rayner and E. Malone (eds.), Vol. 2., Battelle, Columbus, OH, pp. 327–392.

Seppälä O. T. (2004). Visionary management in water services: Reform and development of institutional frameworks. Doct. diss. no 457, Tampere University of Technology, Tampere.

Sharkey W. W. (1982). The theory of natural monopoly. Cambridge University Press, Cambridge; New York.

van der Steen P. and Howe C. (2009). Managing water in the city of the future; strategic planning and science. *Rev. Environ. Sci. Biotechnol.*, **8**, 115–120.

Takala A., Arvonen V., Katko T., Pietilä P. and Åkerman M. (2011). Evolving role of water co-operatives in Finland – Lesson learnt? *Int. J. Co-op. Manage.*, **5**(2), 11–19.

United Nations Development Project (UNDP). (1984). Rural Water Supply Handpumps Project. Laboratory Testing of Handpumps for Developing Countries: Final Technical Report.

Chapter 3

Integration of water and wastewater utilities[1]

Tapio S. Katko, Vuokko O. Kurki, Petri S. Juuti, Riikka P. Rajala and Osmo T. Seppälä

3.1 INTRODUCTION

Integrated Water Resources Management (IWRM) is an idea for international water management and policy that has been strongly advocated for the past decade. Global Water Partnership (2000) defined IWRM as "a process which promotes the coordinated development and management of water, land, and related resources, in order to maximize the resultant economic and social welfare in an equitable manner without compromising the sustainability of vital ecosystems." However, the concept isn't new. Grigg (2008) dates the beginning of IWRM to as early as 1917, when the US Flood Control Act called for "a comprehensive study of the watershed." According to Biswas (2004), the concept of IWRM has been around for as long as 60 years, and it was rediscovered in the 1990s. In the 1970s Finland, Sweden, the United States, and other western countries introduced so-called multipurpose water use and comprehensive water resource planning that were based on many similar ideas.

Although the concepts comprising IWRM are familiar, the idea as a whole has not been fully embraced. This may in part be caused by the abstract language and nature of IWRM as well as its similarities to concepts such as total water management (TWM). Biswas (2004) noted that there were as many as 35 different ways to interpret IWRM. Grigg (2008) observed that "… while *integrated* emphasizes blending together, *total* sweeps in the concept of comprehensive as well as integrated." He further argues that "… unless we use precise terms, each group goes back to the drawing board to create another definition," and therefore there is confusion among IWRM and related concepts.

Saravanan (2006) called attention to the important combination of formal and informal mechanisms of IWRM. A comparative study on the evolution of water and sewage services in 29 cities of 13 European countries demonstrated the variety of administrative and legal traditions. One of the key findings was that merging water and sewage utilities is uncommon in Europe. Of the 13 European countries studied, IWRM is only practiced in Finland and Sweden. Today there is renewed interest in understanding how the integration of service is actually implemented (Juuti & Katko, 2005).

---

[1]Reprinted with the permission of J AWWA.

3.2 MAKING THE CASE FOR INTEGRATION

This article highlights one aspect of IWRM—the integration of urban water and sewage utilities—that has received little attention. A comprehensive overview is given of the issues that need to be considered in a potential merger of these two services by (1) exploring some of the definitions and interpretations of IWRM in the literature, (2) describing how the idea of IWRM was introduced to Finland in the 1970s in the form of comprehensive and multipurpose planning and use of water resources, (3) describing how the idea of integration was incorporated into the management philosophy of community water supply and sanitation services in the 1950s and then in the 1970s was promoted through legislation, and (4) exploring a case study based on the experiences of 14 experts who were involved in the integration of one Finnish city's water and wastewater utilities. It is on the basis of this case study that the authors make some general conclusions (Flyvbjerg, 2004).

A schematic of the integration of water supply and sewage systems is shown in Figure 3.1. The authors have not included stormwater management in this study, although it also is a rising issue in Finland. Finally, the article discusses some of the experiences and drawbacks related to such integration of water and sewage utilities in Finland and other countries.

3.3 INTEGRATION OF WATER AND WASTEWATER UTILITIES IN FINLAND

Viewed internationally, Finnish water and sewage utilities are small, a reflection of the country's population of 5.3 million (The World Bank, 2010). About 90% of Finland's population is served by public water supply, and nearly 80% is served by public sewage systems. Public water service in the country's sparsely populated areas is rather limited compared with many other European countries because of long distances and the abundance of water. However, the number of people receiving public water services has continuously increased, with nearly 100% of the people in densely populated areas receiving services (Hukka & Seppälä, 2004).

3.3.1 Early signs of institutional framework to come

One early sign of the trend toward merging water and wastewater utilities dates back to 1953. That year, the Union of Finnish Cities requested comments on a proposal regarding a special sewage handbook. The Association of Soil and Water Construction Engineers (now the Association of Finnish Civil Engineers) suggested that the handbook also contain information about water utilities. As a result, later that year a book on water supply and sanitation was published—the first of its kind. Since its publication, the single Finnish world "vesihuolto" has referred to both community water supply and sanitation. The shift in thinking that this word represents was likely a precursor to later mergers (Mussalo, 1989). In contrast, most other languages use at least two terms to describe water and sanitation services, for example, "vatten och avloppsvatten" in Swedish, "Wasser und Abwasser" in German, and "servicios de agua y saneamiento comunitarios" in Spanish.

In Finland, the concept of modern community water supply caught hold after World War II, particularly in the 1960s, whereas the concepts associated with modern water pollution control and wastewater treatment expanded most rapidly in the 1970s. In Finland, several administrative and legislative reforms promoted these activities and paved the way for later integration of water supply and sewage utilities in urban centers.

In the 1960s, Finland's National Board of Roads and Waterways was assigned the task of planning a regional water supply system for the Helsinki metropolitan area and for the Turku Region on Finland's southwestern coast. At about the same time, Finland's National Board of Agriculture began preparing regional master plans for water supply and water pollution control in various parts of the country (Katko, 1996; Erävuori, 1976).

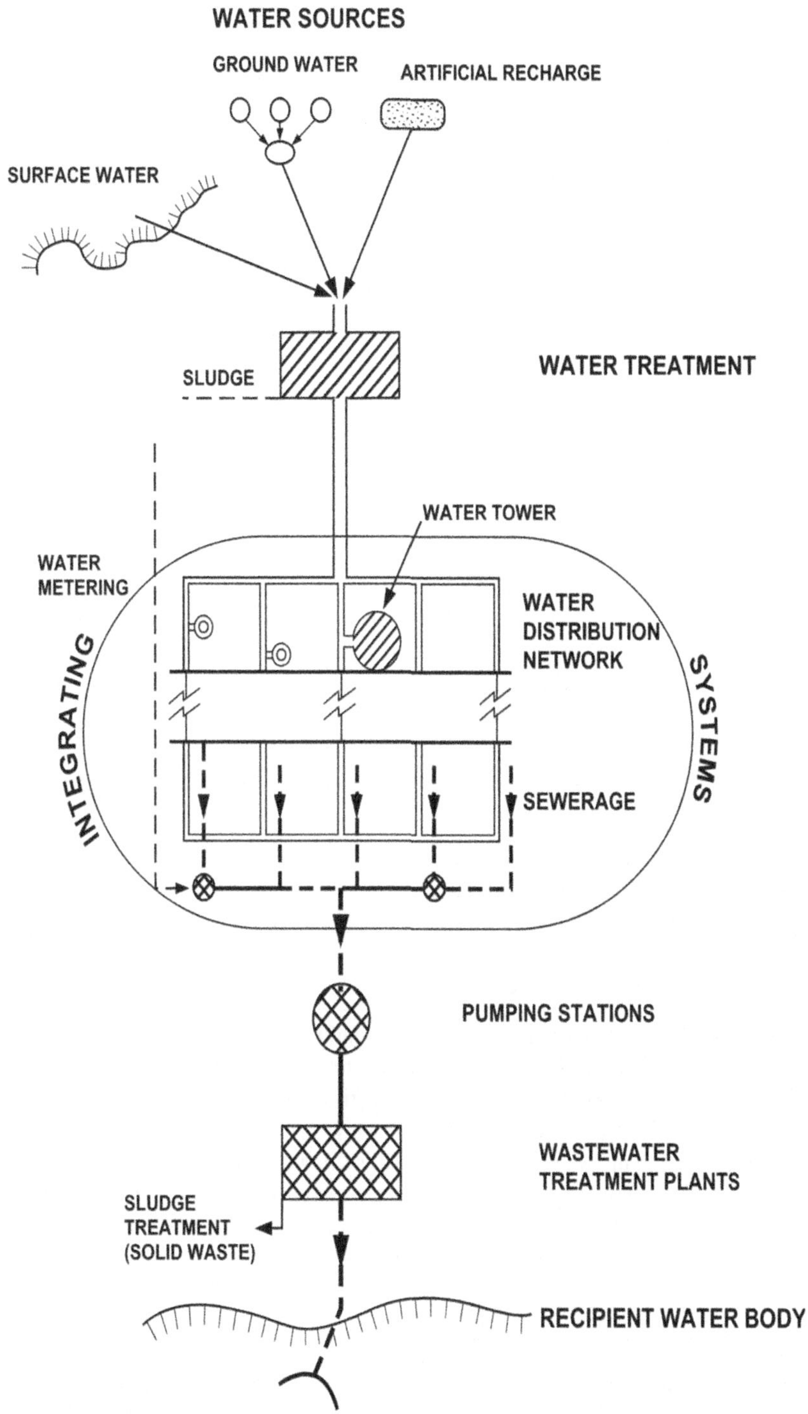

Figure 3.1 Scematic of integrated water supply and sewage systems.

The Finnish Water Administration and the National Board of Waters were established in 1970. Their key duties were promoting the use of, protection of, and research about water resources. Emphasis was placed on comprehensive planning that takes into account the multiple uses of water resources, water pollution control, water supply and sanitation, recreational use of watersheds, the use of hydropower, and flood control (Vesihallinto, 1980). Water supply and sanitation were among the top priorities, as detailed in the comprehensive water resources development master plans put forth by the Finnish administration (Peltokangas, 1996).

For the purpose of comprehensive water planning, the country was divided into 19 areas that corresponded largely with catchment areas, as opposed to water districts that primarily follow administrative boundaries. Regional master plans had previously been prepared for water supply and sanitation, particularly in Ostrobothnia on the western coast where the need for such plans was considered high. Similar master plans were later prepared by consulting companies that were supervised by water and environmental authorities (Katko, 1996). Thus, sectoral master plans were prepared for the most important use—water supply—concurrently with comprehensive water resources development.

Water resources master plans have also served as guides for government support of water supply and sanitation. Individual municipalities were previously quite reluctant to accept these plans, but because of financial trouble that occurred in the 1980s and 1990s, they became more interested in such regional efforts (Katko, 1996).

Water pollution control policies have traditionally been based on long-term-strategies. In 1974 the first water protection program—a national strategy plan for the coming decade—was completed. It was revised in 1985 and again in 1995. These programs identified targets, measures, and instruments. The last one covered the period up to 2005. It also set certain targets for water pollution control by the forest industries as well as all other major polluters (MOE, 1989, 1998).

In 2006 the government adopted a new set of national water protection policy outlines that defined the measures needed to improve water quality. These outlines are in effect until 2015 (Finnish Government, 2006) and define the needs and objectives aimed at

- Reducing the nutrient loads that cause eutrophication;
- Reducing the risks caused by hazardous substances;
- Protecting groundwater;
- Protecting aquatic biodiversity;
- Restoring ecologically damaged water bodies.

In 1995 Finland became a member of the European Union, and thus the European Union Water Framework Directive, enacted in 2000, set additional guidelines for future water management policy. Related to this, in 2004 Finland adopted and started to the organize river basin management planning. The objective of river basin management plans is to improve the quality of surface water and groundwater by the end of 2015. The first river basin management plans were completed earlier this year.

The integration of water and sewage utilities is to be viewed in the context of the wider institutional framework previously described. It covers regional master plans for water resources use, master plans for water supply and sanitation, and water protection programs.

3.3.2 A look at integration-related legislation

The Water Rights Act of 1902 emphasized economic water use and, in particular, promoted hydropower construction. Although the act contained some prohibitions against damming, diverting, and polluting of water courses, water pollution control remained voluntary (Hallberg, 2002).

A few cities started wastewater treatment as early as 1910, but the real boom in modern wastewater treatment facility construction occurred in the 1960s and 1970s. This intensification of construction activity was the result of the water act that went into effect in 1962. It was the first law that introduced discharge permits and had the authority to require communities and industries to begin modern wastewater treatment. Separate sewers began to be introduced at this time, gradually making it technically feasible to treat wastewater.

Two major acts have been of particular importance with regard to the integration of water supply and sewage works. The Wastewater Surcharge Act (WSA) of 1974 more than doubled water rates and replaced earlier systems funded by municipal taxes. This started the debate about which type of utility should operate sewage services (Korhonen, 2010). The WSA and the energy crisis that arrived a short time earlier affected water companies, technology developers, and consumers, led to a decrease in the use of water per capita and in communities, and resulted in more sustainable water use.

The 1977 Act on Public Water and Sewerage Systems in Finland stated that "a Public Water and Waste Water Works is one owned by a municipality or a federation of municipalities accepted by a municipality as such, which has been assigned the task of taking care of a community's water supply and sewerage." This act promoted the integration of water supply and sewage services because they were considered to be naturally linked through the hydrologic cycle. Stormwater was already considered the responsibility of the sewage works, although the appropriateness of this and the ways in which related costs are covered have been debated ever since. Water use and wastewater production can be metered, but stormwater cannot (Korhonen, 2010). The WSA of 2007 currently under revision is expected to recognize stormwater as the responsibility of municipalities, which could buy related services, for example, from integrated water and wastewater utilities. It is hoped that such an arrangement would also solve the problem of how to pay for stormwater management.

3.3.3 Mergers of water and wastewater utilities 1970–2009

Table 3.1 summarizes the mergers of Finland's 20 largest water and wastewater utilities and their related institutional arrangements. Sixteen of them (80%) were citywide water and wastewater utilities, two were regional water and wastewater companies providing both services in their area, and the remaining two had other separate arrangements.

Table 3.1 Mergers of water and wastewater utilities in 20 biggest Finnish cities in terms of population, 1970–1993, reviewed in 2009.

20 biggest utilities*	Year of merger	Notes
Helsinki Water **	1984	In 2010 Helsinki, Espoo and Vantaa established a Joint Regional Authority
Espoo Water**	1974	
Vantaa Water**	X	Established in 1957
Tampere Water**	1981	
Turku Water Works	1989	Wholesale company for raw water; regional wholesale company for WWT since 2002

(*Continued*)

Table 3.1 Mergers of water and wastewater utilities in 20 biggest Finnish cities in terms of population, 1970–1993, reviewed in 2009 (*Continued*).

20 biggest utilities*	Year of merger	Notes
Oulu Water**	1993	
Jyväskylä	N	Water under energy since 2006; Wholesale company for WW since 1971
Lahti Aqua Oy	1970	Divided into 2 companies in 2007
Kuopio Water**	X	A sort of merger in the 1980s, though separate annual reports
Kouvola Water**	1978	Expanded, merger of municipalities in 2009
Pori Water**	1987	
Joensuu Water**	X	Established in 1927
Lappeenranta Water Ltd	1974	Ltd. since 2007
Hämeenlinna Regional Water and Sewage Company	1992	Regional company in 2001: merged with utilities of 6 neighbouring municipalities
Arctic water**, Rovaniemi	1974	
Vaasa Water**	1975	*
Seinäjoki Water**	X	*
Kymi Water Ltd.	2007	Regional water and sewage company of 3 municipalities since 2007
Kotka Water and Sewage Dept.	1977	
Mikkeli Water Works***	1976	*
Porvoo Water**	1975	

In some cases, as in Jyväskylä, two wholesale systems operated in slightly different areas. In Lahti, again, the citywide water and wastewater utility was divided into two municipally owned companies. The motivation in the latter case was to lower or avoid taxes paid by utilities. The energy company in Jyväskylä resorted to "creative accounting" by buying the water utility, not primarily to develop water services, but rather to lower its own taxes (Vinnari & Näsi, 2008).

The numerous mergers of municipalities in Finland in the early 2000s often resulted in only one integrated utility providing water supply and sewage services for the new entity. The three biggest utilities in the Helsinki metropolitan area were merged into a regional environmental services authority for water, sewage, and solid wastes at the beginning of 2010. Although changes occur continuously in organizational arrangements and the trend toward larger systems is not stable, the principle of merged water and wastewater services is largely adhered to in water sales and service distribution.

In the latter part of 1999 through the beginning of 2001, the names of municipal water and sewage utilities were shortened in many cases. From the practical and promotional points of view, this was justified.

However, there is the risk that citizens will not understand the integrated nature of water and wastewater, as was the case previously then the combined term "vesihuolto"—meaning both of these services—or the term "water and wastewater works" was used.

3.3.4 The Pori merger experience

In 1987 the water and sewage utilities of Pori, a city on the western coast of Finland, were merged. The two entities had operated under the governance of two boards. Starting in 1987, the integrated water and wastewater utilities operated under the Board of Construction and Real Estate. In 1998, the utility was turned into an autonomous municipal utility. In 2009, 14 of the utility's staff members were interviewed about the effects of the merger and their related experiences. Their stories are presented in the following sections (Juuti *et al.* 2010). Although the annual utility reports mention the merger in 1987, there is no discussion of the integration process in subsequent years.

Before the merger, working groups for planning, construction, networks, treatment plants, customer service, technical support service, and administration and management were established for planning the integration and related actions (Soukki, 2010). The groups were divided further into teams to look into more specific issues, for example, the teams of the customer service group studied issues related to supervision, home connections and metering, and customer agreements (Hedberg, 2010).

The first water/wastewater integration proposal had actually been presented in 1974. It took 12 years to finally implement the merger after four failed attempts with slightly different setups (Soukki, 2010). When the new director of the water utility was hired in August 1984, the merger became one of his major responsibilities. Because he had not previously been employed by either the water or sewer utilities, he was able to operate quite independently.

The water utility strongly opposed becoming merged under the Board of Construction and Real Estate, and the staff did not want to be merged with the wastewater staff—something they regarded as a demotion. After the merger, the work climate at first deteriorated because it was commonly felt that the integration had been forced. A clear conflict between the working cultures existed: the water utiity staff was unwilling to work with the sewage utility staff. Those who considered themselves water utility experts claimed that they would have nothing to do with sewers. The water utility employees had adjusted to the idea that costs were paid for by consumer fees, but the older practice of funding sewers with taxes, which had been in place before the 1974 Wastewater Charge Act, was also still in use. Various opinions were expressed about the amount of time that would be needed to integrate these two work cultures— in reality, it took a decade.

On the positive side, it was noted that assistance from the sewage utility was available to the water utility when needed. After the merger, pipe laying and other related activities were planned and implemented jointly. Operations became more logical and efficient and often two pipe layers could do the work of four. In many respects, the integration was one of the biggest changes in water services management in Pori—although the change was resisted, advantages were realized and accepted over time.

3.3.5 Other findings from Finland

According to Korhonen (2010), the following economic advantages in particular can be reached by merging water supply and sewage services:

- At least part of the staff can work for both services;
- Cost savings from laying water and wastewater pipe in the same trench—as done in Finland—can easily be allocated according to actual expenses.

A recent study on the options for and experiences with managing intermunicipal water supply and sewage systems in Finland (Pietilä & Katko, 2010) implies that the raw water sources and natural and artificial groundwater intake areas are often geographically distant from larger wastewater collection and treatment points. This favors the principle of flexibility and need to take into account local conditions, including various options for wholesale arrangements, if needed.

Korhonen (2010) refers to the oldest wholesale company for regional water supply that was established in the Kalajoki River Valley in 1968 (Hannula, 2008). Since 2008, the company has also provided wholesale wastewater treatment—thus it is an integrated wholesale company. According to Korhonen, in the case of the 110-km long Kalajoki valley it would not be wise to merge all the municipal water distribution and sewage utilities into a large, comprehensive regional company, but local offices would be needed for on-call services.

In Finland's capital, Helsinki, water and wastewater utilities were merged in 1984. According to Tiainen (2009), before the merger the water utility employed 450 people and the sewage works employed 270. The merger resulted in a reassessment of the activities of the new utility. For example, the water utility had its own engineering works, employing 30 people who manufactured equipment such as water hydrants, which at the time could have been purchased externally for less. This practice was unique in Finland. The old utility also had an office that employed more than 30 people who built water connections for other city-owned utilities, hospitals, and homes for he elderly. This activity, commonly practiced by many urban utilities, ended in 1988 in Helsinki. By 1995 the staff of the merged utility had fallen to fewer than 400 employees (Herranen, 2002) and by 2008 it was closer to 300 (Helsinki Water, 2008).

In the case of Helsinki, the merger of the water and sewage utilities was initiated in 1975 when the city technical works committee proposed studying it. At that time the water utility was in favor of integration because it considered the chain of operations comprising both services to be both a technical and economic entity. But the public works department, which was in charge of sewage, held the opposite view. The only advantage from integration, according to the public works department, was that both utilities could lay pipes in the same excavation. Besides, in areas that had combined sewer systems, the sewers were an integral part of street construction (Herranen, 2002).

Findings from several case studies indicate that staff members of water supply utilities often looked down on staff members of sewage utilities—they were considered to be involved in something dirty and less valuable. However, among members of the general public attitudes seem to be changing. Because of public relations activities, people have recently started showing interest in understanding how their wastewater is treated and managed.

3.4 THE IWRM EXPERIENCE IN OTHER COUNRIES

According to Persson (2010), the Swedish Public Water and Wastewater Plant Act of 1970 determined the responsibilities of the water and sanitation supplier and customer, but it did not specifically side with larger supply systems. The Danish Water and Waste Water Association has 152 utility companies as members. Of these, 43 (29%) provide water and wastewater services, 63 (41%) provide water only, and 46 (30%) provide only wastewater services. According to Fischer (2010), it is likely that some mergers of water and wastewater services will take place in five to 10 years. To a degree this has already begun. In connection with Denmark's latest reform there has been a consolidation of the sector, corresponding with a reduction of municipalities from 270 to 98.

In the mid-1800s most western nations began developing urban water and sewage services based on private concessions or operators, but fairly soon the operations were taken over by municipalities. Only in France have private operators survived and expanded considerably. This is largely because France has so many municipalities—36,000 in 2000. It is difficult to imagine so many municipally owned utilities all managing their own water services (Juuti *et al.* 2005).

With regard to municipal hierarchy, some interesting traditions exist in Europe. For example, the "Stadtwerke" in Germany operates both water and gas, but not sewage systems. A special feature of the Dutch water sector is the water boards—independent decentralized governmental entities with elected members. These boards oversee wastewater treatment, whereas water supply and sewage are typically separate utilities or bodiesunder municipalities (Pietilä, 2006).

Merged water and wastewater utilities certainly also exist in other countries, but it is difficult to find reliable statistics about them. Because water and sanitation services are typically managed at the local level, the role of local governments and legal and administrative traditions largely determine how these services are managed in each country (Juuti & Katko, 2005).

In 1996 AWWA surveyed 898 US utilities serving from 2000 to 5 million people. Of the 881 utilities that responded, 351 (40%) had merged their wastewater treatment and water supply. In a subsequent survey of distribution systems, AWWA (2003) received data from 339 utilities serving from 3000 to 4 million people; of these, 151 (45%) also provided wastewater treatment. It appears that the number of utilities providing both water supply and sewage service is increasing. Additional studies are needed to verify this speculation.

A case study by Isbell & Lee (2006) examined the integration of water supply and sewage utilities in Charlotte, N.C., in 2005. The approach taken by the utilities was to focus on managing individual strengths and lacing the right people in the right roles. A pilot program indicated that costs could be reduced by

- Creating crews of mixed skills;
- Consolidating headquarters;
- Cross-training drinking water and wastewater personnel;
- Increasing the responsibilities of team leaders;
- Sizing crews properly;
- Enhancing communication.

Isbell and Lee concluded that integration is an ongoing process and the management challenge is to determine individuals' unique talents and then bring them together.

In countries with developing economies (e.g. Accra in Ghana), it is not uncommon that international financial bodies have promoted the idea of private operators developing water supply services, whereas sewage and wastewater treatment are provided by other organizations. Before the water sector reform during 1993–94, the water supply and sanitation services were integrated under one national entity, the Ghana Water and Sewerage Corporation, and served approximately 370,000 water connections and 6–8 million people in the country's 10 political regions. Although water supply remained centralized, the responsibility for sanitation and wastewater management was decentralized (Suleiman & Cars, 2010). With this organizational model, there is the risk that the lack of integration will make it difficult to implement effective water pollution controls and ensure that the system is financially sound—an acute challenge for many developing and transitioning economies.

Moriarty and colleagues (2010) suggest that in Middle Eastern countries instead of the typical top-down IWRM structure, these countries should instead apply so called light IWRM; in other words, focus on delivering water- based services to people. However, according to Moriarty *et al.* a major limitation of this approach is the lack of appropriately decentralized finance. Local authorities typically rely on financing from the national government, which is often earmarked and over which they usually have little control.

3.5 WHAT EXPERIENCE CAN TEACH US

When considering integration of water supply and sewage utilities in the context of Finland, it is important to recognize the four major system types. At the lowest level, there are onsite water and sanitation systems. These typically serve one or a few households and are not connected to networks. The next level is small rural systems that serve villages or a slightly larger area. These systems are commonly managed by cooperatives and increasingly are becoming involved with sewage services. The third level of systems is those serving townships and cities and surrounding population centers that form a single municipality. Fourth, there are a variety of intermunicipal systems, most of which operate on a wholesale basis both in water supply and sewage service. Integration of water supply and sewage utilities in Finland occurs mainly on the third level. However, the small water cooperatives of rural areas are also increasingly expanding to include sewage services because of the decree that requires adequate water pollution control in the case of permanent and secondary housing in rural areas by 2014. Some water cooperatives are also likely to merge with larger municipal or other types of systems. The overall structure is close to that of Sweden except that their municipally owned systems are responsible for smaller systems.

As for the Pori case, through extensive interviews Sandelin (2006) found that although the two utilities had merged in 1987, almost two decades later several cultures still prevailed within the utility because of geographical distance. This is another possible drawback when considering the ever-expanding size of water and sewage utilities that is now somewhat indiscriminately promoted by authorities in Finland.

In one recent case, there was strong opposition to merging the oncall duties of water and wastewater system operators. It is true that any possible contamination risks must be considered carefully. However, as Rontu (2010) points out, it is probably more a question of performing the duties in the right order; water supply should get preference.

The following conclusions were drawn concerning the merging of water and wastewater utilities in Finland.

(1) Merging water and wastewater utilities under one utility or organization seems to have several advantages, at least for larger urban areas and retail activities.
(2) In a wider regional and intermunicipal context, merging is probably more complicated and less feasible.
(3) In the case of small systems, more mergers between piped water and wastewater services and with bigger systems can be expected.
(4) Preparation of mergers as well as the development of joint management cultures after integration will take time, easily a decade in each case.

Yet, integration of water supply and sewage services is logical based on the experiences of Finland and several other countries. From the perspective of people—the primary users of water—ensuring adequate cooperation between water and sewage utilities could be the first natural step toward more integrated water resources management. More studies on the experiences of integration of water supply and sewage services, including their advantages and limitations in various conditions, would be useful in moving the field forward.

Acknowledgement

The authors thank Jorma Tiainen for assistance with language, Neil S. Grigg for his advice, as well as Kaj W. Hedberg and Ilkka Mikkola for additional assistance. Additionally, the comments by the editor and the three peer reviewers are highly appreciated. The views of the 14 experts from Pori, Finland, are also greatly appreciated, and the financial support from Pori Water and Academy of Finland (decision number 135843) is gratefully acknowledged.

REFERENCES

AWWA (2003). Waterstats 2002, Water Utility Database. AWWA, Denver.

AWWA (1996). Waterstats 1996, Water Utility Database. AWWA, Denver.

Biswas A. K. (2004). Integrated water resources management: a reassessment. A water forum contribution. *Water Intl.*, **29**(2), 248–256.

Erävuori J. (1976). *100 vuotta vesilaitostoimintaa Helsingissä* (Water Supply Operations in Helsinki). Helsingin Kaupungin Julkaisuja. 28. Helsinki.

Finnish Government (2006). Water Protection Targets for 2015. www.ymparisto.fi/default.asp?contentid=210839&lan=en (accessed 22 April 2010)

Fischer L. (2010). Danish Water and Waste Water Association. Personal communication. 26 April 2010.

Flyvbjerg B. (2004). Five misunderstandings about case-study research. In: C. Seale, G. Gobo, J. F. Gubrium and D. Silverman (eds.), Sage Publications, London.

Grigg N. S. (2008). Total Water Management: Practices for a Sustainable Future. AWWA, Denver.

GWPTAC (Global Water Partnership Technical Advisory Committee) (2000). Integrated Water Resources Management. TAC Background Papers. GWP Secretariat, Stockholm. www.cepis.ops-oms. org/bvsarg/i/fulltext/tac4/tac4.pdf (accessed 22 April 2010)

Hallberg P. (2002). Vesioikeuslaki 100 vuotta (Water Rights Act 100 years). *Vesitalous*, **43**(5), 7–8.

Hannula J. (2008). *Vesikolmio 40 vuotta. Alueellisen vesihuollon edelläkävijä.* (Vesikolmio 40 years. Forerunner for Regional Cooperation), Ylivieska.

Hedberg K. (2010). Personal communication. 4 Mars 2010.

Helsinki Water (2008). *Helsingin Vesi* (Helsinki Water). http://wapedia.mobi/fi/Helsingin_Vesi (accessed 22 April 2010)

Herranen T. (2002). 125 Years of Life with Water. The History of Water Services in Helsinki. Helsinki Water, Helsinki.

Hukka J. and Seppälä O. (2004). National Context Report. WaterTime, Finland. www.watertime.net/wt_cs_cit_ncr. asp#Finland (accessed 22 April 2010)

Isbell M. and Lee A. (2006). Charlotte-mecklenburg successfully integrates field operations by managing individual and organizational strengths. *J. AWWA*, **98**(9), 85–90.

Juuti P. and Katko T. (eds.) (2005). Water, Time and European Cities. History Matters for the Futures. European Commission, EU. www.watertime.net/Docs/WP3/WTEC.pdf (accessed 22 april 2010)

Juuti P., Rajala R. and Katko T. (2010). *Vettä eks tiä? Näkymätönt Porrii. Porin veden historia* (History of Water Services in Pori). Pori Water Pori.

Katko T. (1996). *Vettä!—Suomen vesihuollon kehitys kaupungeissa ja maaseudulla* (Water! — Evolution of Urban and Rural Water Supply and Sanitation in Finland). Finnish Water and Waste Water Works Assn., Helsinki.

Korhonen M. (2010). Personal communication. 26 February 2010.

MOE (Ministry of the Environment), (1989). Water Protection Programme to 1995. Decision-in-Principle by the Finnish Council of State. Series B, number 14.

MOE. (1998). *Vesiensuojelun tavoitteet vuoteen 2005* (Goals for Water Pollution Control by 2005). *Suomen Ympäristö*, **226**, 82 p.

Moriarty P. B., Batchelor C. H., Laban P. and Fahmy H. (2010). Developing a practical approach to 'Light IWRM' in the Middle East. *Water Altern.*, **3**(1), 122–136.

Mussalo R. (1989). Personal communication.

Peltokangas J. (1996). Personal communication. 10 June 1996.

Persson K. M. (2010). Personal communication. 2 Mars 2010.

Pietilä P. (2006). Role of Municipalities in Water Services. PhD thesis, Tampere University of Technology, Finland.

Pietilä P., Katko T. and Kurki V. (2010). *Vesihuolto kuntayhteistyön voiteluaineena*, (Water Flueling Municipal Collaboration). Foundation for Municipal Development.

Rontu M. (2010). Personal communication. 15 January 2010.

Sandelin S. (2006). Knowledge Management and Networking in a Water Utility – The Case of Pori Water. Licentiatés thesis, Tampere University of Technology, Finland.

Saravanan V. S. (2006). Integrated Water Resource Management. A response. *Economic and Political Weekly*, 41(Sept. 23). www.zef.de/module/register/media/18cf_Saravanan,%202006-EPW-41-38.pdf (accessed 22 April 2010)

Soukki R. (2010). Personal communication.

Suleiman L. and Cars G. (2010). Water supply governance in accra: 'Authentic' or 'Symbolic'? *Water Policy*, **12**, 272.

The World Bank (2008). The World Bank Group, Finland. http://data.worldbank.org/country/finland (accessed 17 August 2010)

Tiainen E. (2009). Personal communication. 2 December 2009.

Vesihallinto (1980). Vesihallinto 1970–1980. (Water Administration), Helsinki.

Vinnari E. M. and Näsi S. (2008). Creative accrual accounting in the public sector. 'Milking' water utilities to balance municipal budget and accounts. *Financ. Account. Manage.*, **24**(2), 97–116.

Section II

Technology and Socio-Ecological Interaction

Section II Technology and Socio-Ecological Interaction consists of five chapters. The 1st chapter was written by two authors, Sauri D. & del Moral L. They apply the historical approach in studying the governance of large hydraulic infrastructures in Spain.

The 2nd chapter was written by one author, Lepper T. He concentrates on Colorado's water development and the pursuit of sustainable growth in the arid west.

The 3rd chapter was written by one author, Fort C. She is a historian who examines the future of South Australian water recycling and sustainability.

The fourth chapter was written by one author, Tempelhoff J. His area of interest is the rising tide of public concern over acid mine water drainage in South Africa.

The last chapter was written by three authors, Vuorinen H. S., Juuti P. S. & Katko T. S. They concentrate on the safety of lead water pipes in history and the present.

Chapter 4

Governance of large hydraulic infrastructure in Spain: a historical approach

David Sauri and Leandro del Moral

4.1 INTRODUCTION

The hydraulic infrastructure built in Spain has created some of the most modified river systems in the world by changing the landscape of many Spanish regions considerably, mainly through irrigation. Since the 16th century, the construction of dams and reservoirs has gone through different stages, including the failed attempts of the Enlightenment and the Liberal periods (18–19th centuries) and continuing with the state taking over the responsibility for building and operating them in the late 19th century. Each period has been characterised by different modes of governance, but in this chapter we will be particularly interested in tracing changes in governance from the early era of historical reservoirs such as Tibi (Alicante), Níjar (Almería) or Puentes and Valdeinfierno (Murcia) to the hegemonic role of the Confederaciones Hidrográficas (Water Basin authorities) since the 1930's and the emergence of new voices and values in water management stimulated by the new territorial configuration of the Spanish state into *Comunidades Autónomas* (regions) and by the European Water Framework Directive of 2000 (Figure 4.1).

The chapter is organised in chronological order. The first part is dedicated to tracing the steps towards the constitution of a national water policy from the late 19th century until the end of the Spanish Civil War (1939). The second part focusses on the Franco period (1939–1975) when the hydraulic paradigm reached probably its highest splendor with the *Generalissimo* inaugurating dams and reservoirs so often that people in secret called him "Frankie the Frog". In the third part we explore the most recent period marked above all by a remarkable shift in water policy and management towards a crisis concerning large water works (including the failure of the inter-basin transfer of the Ebro), the emergence of new forms of water governance based on demand management and alternative resources, and the reconfiguration of the hydraulic paradigm into new, more subdued forms such as desalination.

Figure 4.1 Administrative divisions (thinner lines) and river basin divisions (thicker lines) in the Iberian Peninsula. Names within rectangles correspond to river basins.
Source: Own elaboration.

4.2 WATER POLICY IN THE ANCIENT REGIME

The beginning of the construction of the modern State of Spain in the sixteenth century coincided with a noticeable expansion of public water works. The main attention focussed on the construction of dams including those in Almansa (in the province of Albacete) and Tibi (in Alicante, south of Valencia) in the Jucar River basin. Both are dams curved in plan with systems for the release of mud that represented a major technical innovation and are still in use. The Elche dam (province of Alicante), which was constructed in the first half of the seventeenth century, is considered by some authors the first vault-dam, that is, a dam curved in plan (arch dam) with uniform thickness from top to the bottom (Fernández Ordoñez 1984; López Gómez 1987, 1992).

Compared with the Spanish water politics of later times, this era constituted an example of fruitful innovation. Norman Smith, perhaps with some exaggeration, assures that: "Spain was the birth-place of modern dam building" (Smith, 1970:32). Antonio López Gómez also writes that: "The 16th and 17th centuries are an extraordinary stage in such works, unparalleled in Europe" (López Gómez, 1992:125). Such statements, often repeated, have helped create and develop a discourse that is a key factor of the institutional and conceptual "third layer" (Tvedt, 2009) of the Spanish water system.

The new Bourbon Dynasty ruling Spain since the 18th century showed a firm commitment to regaining Spain's role in the international arena. The objective of unifying Spain by means of a suitable system of communications gave river navigation a central role. However, the Enlightenment misinterpreted the country's physical features, mainly the high average altitude and rough orography, and the serious difficulties created by the fluvial regime, that is, summer rainfall shortages and consequent acute low summer flows of Spanish rivers. Throughout the century, the effective fluvial transport schemes implemented in other European countries were a model for Spanish elites who fell into a mirage caused by the (erroneous) extrapolation concerning the possibility of constructing a similar extensive channel network in Iberia.

Along with (failed) navigation projects, the hydraulic works of greatest impact in the 18th century were the Puentes and Valdeinfierno dams, on the River Guadalentín, both representing a milestone in European hydraulic history. In contrast with the 5.4 million cubic metres of the Tibi Reservoir, the largest existing dam until then, the capacities projected for the new dams rose to 29.5 (Valdeinfierno) and 52.0 million cubic metres (Puentes). After eleven years of operation, on the 30th of April 1802, the Puentes dam collapsed and caused the biggest disaster in Spanish water history with 608 dead. The tragedy of the Puentes dam stirred a European controversy about reservoirs and their safety that lasted until the beginning of the 20th century. Despite these failures, important achievements were made in the 18th century in the form of several completed projects (Arroyo & Camarero, 1989). The ones worth mentioning are the Aragón Imperial Channel and the Tauste channel, both in the River Ebro; the Castile and the Campos channels in the Duero river basin; the restoration of the Gran Prior channel in the Guadiana, and the lengthening of the Alcira Royal Channel (*Real Acequia de Alcira*, originally built in the thirteenth century in the Júcar river basin).

4.3 REGENERATIONISM AND CONSOLIDATION OF THE HYDRAULIC PARADIGM

At the end of the eighteenth century, the priority of water politics turned shifted from navigation to irrigation. In his *Informe en el Expediente de la Ley Agraria* (Report on the Agrarian Act), published by the Economic Society of Madrid in 1795, Jovellanos, one of the most prominent exponents of the Enlightenment reformist policies in Spain, already noticed that the expansion of irrigated land in Spain was both urgent and difficult. In 1820, the *Ley sobre Caminos y Canales del Reino* (the Kingdom's Roads and Channels Act) created a parliamentary Commission on this topic and initiated a discourse that would become obsessive during the following decades: "Not a single drop of water will go to the Ocean without paying due tribute to the Earth, because it takes with it a great part of what constitutes our wealth, exposing our laziness and lack of responsibility". The Commission openly defended the national character of the main canals to be constructed, and therefore, the pivotal role to be played by the central Government in this regard. Thus, the 1820 proposal suggested an idea which the hydraulic *Regeneracionismo* would also defend with vehemence at the end of the century. However, the proposal of the 1820 Commission was not accepted during the rest of the 19th century. On the contrary, several legislative measures were passed to facilitate the private transformation of dry land into irrigated land. None of these norms worked properly, and the attempts to entice investment in water infrastructures were largely unsuccessful (Del Moral, 2010).

In 1898, whilst other industrialised European countries were consolidating their colonial conquests, Spain was in a state of shock at the loss of its last colonial possessions (Puerto Rico, Cuba and the Philippines), following a disastrous war against the United States. In the absence of an overseas colonial project as a means for modernisation, Spanish elites advocating social and economic reforms (the so-called *regenerationists*) concentrated their efforts on a national programme involving a radical transformation of the country's geography. This vision combined a firm political strategy, a call for a scientific-positivist understanding of the natural world, a scientific-technocratic engineering mission, and a populist base rooted in the traditional peasant culture. *Regeneracionismo* thus united diverse social and political sectors (reformists, socialists, populists, industrialists and agricultural elites), while keeping the more radical left-wing forces (revolutionary socialists and anarchists) and the traditionalist conservative right at bay. Furthermore, water politics played an important role in the social legitimisation of the State. Reformism advocated *hydraulic regeneration*, whereby the State would take centre stage in organising the necessary physical and socio-economic transformation of the water resources of Spain (Del Moral, 2010).

Within this context, the "hydraulic mission" became the central idea of the *Regenerationism.* This mission was conceived as a strategy to remedy the national economic and social problems and to redeem ("re-generate") Spain's troubled geography. There was basic agreement among the regenerationists that "the evils of the mother country" have, to a great extent, a physical foundation, due to the absolutely unfavourable conditions of relief and climate" – for example, Joaquin Costa, Lucas Mallada, Macías Picavea and Torres Campos y Reparaz were of this opinion (J. Gómez Mendoza, 1992:231–235). According to Costa: "… the central plain, and perhaps half of Spain, is one of the driest regions of the globe, after the deserts of Africa and Asia (…) the atmospheric currents of the Mediterranean and the Atlantic do not spill on the scorched fields of the Peninsula all the water which the plants need to vegetate and to fructify; but there are immense water deposits in the crests and entrails of mounts, and we can spill it over the country with the mathematical regularity of the pulsations, crossing it with a hydraulic arterial system that mitigates its heat and extinguishes its thirst …" (Costa, 1880; cited in Gil Olcina, 2002:16).

Because of the limited prospects of private investment, the regenerationists saw the central State as the only body capable of mobilising the funds required to develop the country's water resources. They pushed through the necessary reforms in the face of strong opposition from traditional oligarchies. The efforts of a "developmental" and modern state were considered vital for re-engineering the country, both socially and environmentally. At the same time, this reformist route was able to secure the support of part of the old elites, as it did not threaten their fundamental rights as landowners and defended rural power against the rising tide of the urban-industrial elites and the proletariat (Saurí & Del Moral, 2001). Thereby the 1911 Hydraulic Works Act recognised the State's active role regarding the direct execution of certain large irrigations and flood defence works. Twenty years later, the Urgent Implementation of Irrigation Act (*Ley de Puesta en RiegoUrgente*) of April 1932 established that the State had to make the necessary complementary works for implementing irrigation in areas already subjected to large hydraulic works (dams and main canals, in some cases publicly funded thanks to the 1911 Act), but still without proper irrigation infrastructure (Ortega Cantero, 1984:122).

The Plan Gasset (1902) was the first attempt to push forward the regenerationist agenda. It was followed by several updates (1909, 1915, 1919, and 1922) that attempted to re-enforce the hydraulic project, although with little success (Gil Olcina, 2002). Another significant milestone of the regenerationist agenda was the establishment of river basin authorities (*Confederaciones Hidrográficas*), probably among the world's first river basin authorities were gradually established between 1926 and 1961. The initial structure of the *Confederaciones* was based on four principles: the river basin as the proper scale for the management of water resources; the water basin as an integrated planning unit; participation of water users in the management of the river basin, and decentralisation of state functions to the river basin level (Swyngedouw, 1999). During the Second Republic, the First National Water Plan (1933) was passed. A key part of it was the proposal to transfer water from Tagus River, which discharges into the Atlantic near Lisbon, to Segura River which flows into the Mediterranean. The idea was to promote the intensive agriculture of the Eastern regions and enlarge the traditional and fertile irrigated areas of Valencia and Murcia, which were already producing competitive export commodities (López Ontiveros, 1998). The outbreak of the Civil War (1936–1939), however, prevented the implementation of the Plan, but its technical and infrastructural proposals – although in a largely watered down version compared to the original regeneracionist platform – were carried out after the Civil War.

4.4 THE GOLDEN ERA OF THE HYDRAULIC PARADIGM

After the victory of Franco in 1939, river regulation through the construction of reservoirs continued unchanged and even intensified compared to previous period. In 1940, the total capacity of Spanish reservoirs was 3914 cubic hectometres. By 1972 it had increased to 36,335 cubic hectometres: almost 10 times the capacity at the end of the war (Nadal, 1993). In 2005, the total estimated capacity of the over 1000 large reservoirs built in Spain had risen to some 56,000 cubic hectometres (Figure 4.2).

Figure 4.2 Melonares Dam (near Seville) under construction. This is a project addressed to supply water to the City of Seville.
Source: Author Leandro Del Moral.

Although the official rhetoric for justifying the development of large reservoirs made irrigation first priority, in keeping with the regenerationist agenda, generation of electricity was what drove Spanish water policy in this period at least until the 1970's. In fact, industrial and urban growth followed in the footsteps of power generation in Spain beginning in the 1960's (Molina & Montiel, 2004). Installed capacity grew from 1350 Mw in 1940 to 18,400 Mw in 2005, with the strongest expansion between the 1950's and the 1970's. About a third of the reservoirs built before the 1980's were specifically designed for hydroelectricity generation with a total installed capacity of 17.9 million kWh. Since Spain lacked other sources of energy (especially oil and good quality coal), the development of hydroelectricity was given a high priority by the Franco regime to the point that in 1960 approx. 85 per cent of the energy produced in Spain came from hydroelectrical plants. Since then, the share of electricity generation has declined sharply to a mere 15–20 per cent (depending on the water amount stored in reservoirs) because of the growing difficulties in finding appropriate sites and, above all, the increase in thermal energy production (coal, oil, gas and nuclear). The largest hydropower plants (*Aldeavila* on the Duero River, and *Jose María de Oriol,* on the Tajo River) were built in the 1960's and 1970's. The close relationship between reservoir construction and electricity production is reflected in the 10-fold increase in hydroelectrical capacity recorded between 1940 and 1970 (A. Gómez Mendoza *et al.* 2007).

Hydroelectrical production joined irrigation and a growing urban and tourist demand as the main purposes of reservoirs, especially in Eastern Spain. Some of the largest projects of the 1960's and 1970's involved inter-basin water transfers. The most important one was the so-called Tagus-Segura transfer, already suggested in the 1930's, which aimed to divert water (some 600 cubic hectometres per year on average, but much less in drought periods) from the Upper Tagus Basin to the burgeoning agricultural and tourist areas of the arid South-East (Figure 4.3).

Figure 4.3 Bolarque Dam in the Tagus River and intake point for the Tagus-Segura aqueduct.
Source: Author Leandro Del Moral.

The Tagus-Segura fitted well into the dominant logic that "water had to go where it can produce the highest economic and social benefits" (Gil Olcina & Rico Amorós, 2008). However, this logic failed to justify another *Trasvase,* that concerning the Ebro River, which in the early 1970's attempted to quench the thirst of Barcelona by transferring some 1000 cubic hectometres per year to the city and its metropolitan area. Strong opposition from the donor basin plus the economic crisis and the political change in the mid- to late 1970's shelved the project which would, however, be resuscitated two decades later.

Governance of large water infrastructures during the hydraulic era was largely conducted by the *Confederaciones Hidrográficas* (River Basin authorities). These organisations adhered to the corporativist model of social relations developed in Fascist Italy. Under this model, the most influential stakeholders in river basins (hydroelectrical companies, irrigation interests, and water supply companies) joined expert public servants (most notably civil engineers) to create basin water management boards with ample decision making powers. It was a classical top-down management model which excluded outsiders. The *Confederaciones* represented thus the techno-managerial approach intended to orient water management towards satisfying the needs of large water users, and especially (at least until the 1970's) hydroelectric power interests (Garrido & Llamas, 2009).

4.5 THE DEMISE OF THE HYDRAULIC PARADIGM AND THE EMERGENCE OF NEW WATER POLICY ALTERNATIVES

The death of Franco in 1975 and the beginning of the democratic era in Spain did not bring a radical change to water planning and management – large reservoirs continued to be built. In fact, the largest reservoir in Spain (la Serena, on the Zújar River, in Extremadura, with a capacity of near 3000 cubic hectometres) was inaugurated in 1990. However, the new democratic winds blowing in the country also gave rise to the first important social protestations against large projects. Some of these projects were cancelled or downsized, especially in the Pyrenees. But growing social opposition (e.g. to Riaño, in the Northwest of Spain, or to Rialb, in the Northeast) did not deter public authorities from continuing with their policy of business as usual. Thus in 1993, the Socialist Party, then in power, presented a water plan called "Integrated National Water Balance" consisting of reservoirs, canals and aqueducts aimed at nothing less than the transfer of 3,350 annual Hm3 from the northern rivers to the thirsty South and East. The Socialist Plan of 1993 followed the logic of the hydraulic paradigm. However, times had changed, and the plan was rapidly shelved mainly because of some (extravagant) demands for irrigation water. In 1999, the conservative Popular Party introduced a number of modifications to the Spanish Water Law of 1985 in the direction of facilitating the creation of water markets and, thus, providing interested parties access to water without resorting to new supply sources. The initiative was received with little enthusiasm, and a new national water plan, more modest than the 1993 project but equally insistent on building new reservoirs and aqueducts, was drafted and approved in 2001, this time by the conservative Popular Party. The main feature of the 2001 Plan was the transfer of some 1.000 cubic hectometres annually from the Ebro River to Valencia, Murcia and Almeria in the East and South, and to Barcelona, in the North. Radical opposition from the Ebro Basin (including mass demonstrations in Madrid, Barcelona and Brussels), the reluctance of the European Commission to provide funds for the project (because of concerns about the environmental fate of the Ebro Delta), and – above all – the Socialist victory in the Spanish national election of 2004 put this project to a rest again. Yet, large hydraulic projects can also be transnational. In the 1990's, the Catalan regional government, and the French region of Languedoc Roussillon agreed to transfer water from the Rhône River south to Barcelona. This project was framed under the banner of the "Europe of the Regions" but high economic costs, the need of a treaty between France and Spain (and not between the two regions), and some obscure financial matters affecting *the Compagnie de Bas Rhône- Languedoc* (responsible for the project) precluded further progress.

Water governance during this period changed in important ways with respect to the previous era. First, in 2000, the European Union issued the Water Framework Directive (WFD) which signaled a substantial departure regarding water planning and management on the Continent. From then on, European water policy would have to pay attention to the ecological status of water, cost recovery principles, and public participation. Closely related to the WFD was the emergence in Spain of a movement called "Nueva Cultura del Agua" (New Water Culture) which called for a radical change in water policy away from large projects causing irreversible environmental and social impacts, such as the Ebro project, and more attention on demand-side solutions and above all public participation. Both the WFD and the New Water Culture influenced greatly the water policy adopted after the Socialist victory of 2004 which bore little resemblance to the party's earlier policy that had approved the massive development of Spanish rivers in 1993. After 2004, Spanish water policy appeared to disfavour large hydraulic works in rivers promoting instead the new alternative: desalination. The so-called "AGUA" Programme thus led to the construction of some 20 desalination plants along the Mediterranean Coast to supply the water that otherwise would have come through the Ebro transfer (Figure 4.4).

Figure 4.4 Main room of the desalination plant of Barcelona.
Source: Author Hug March.

The call to public participation in water affairs issued by the WFD also induced changes in the governance of river basins. Management boards had to include not only traditional stakeholders but also new members from environmentalist groups, social movements, unions, academia, and so on. Beyond this, some regional water agencies (equivalent to river basin authorities) organised public meetings open to all interested citizens for the formulation of proposals for better water management practices. The Catalan Water Agency, for example, asked users and the public in general about their priorities for river management in several basins of Northern and Eastern Catalonia. However, the critical voices on river basin management boards were often outnumbered by the technical staff of river basin authorities and by traditional stakeholders, and public proposals often clashed with the realities of current economic and political arrangements. In any case, this period meant an important expansion of water governance in Spain, although traditional vested interests still control decision making to a large extent.

Finally, the territorial configuration of the Spanish State after the Constitution of 1978 added a new layer of complexity to water planning and management with important implications for water governance. According to the Constitution of 1978, Spain had to be divided in 17 *Comunidades Autónomas* (regions) endowed with important legislative and executive powers in several areas of public affairs. As to water, each *Comunidad Autónoma* would have exclusive control of the water resources of watercourses flowing entirely within its administrative limits. If, as is often the case, river basins are shared by more than one region, the political body responsible for water planning and management is the Spanish state through River Basin Authorities. This constitutional agreement has been interpreted, however, in different ways by various regions, especially as regards the sharing of water flows and the possibility of transferring water to other regions. Mostly in order to have determining control over water flows, many *Comunidades Autonomas* have attempted to have the last word in deciding the final destination of the surface waters existing in their territories. The region of Aragon, covering a substantial part of the Ebro Basin, decided, for instance, to prohibit any water transfers outside its limits before the supposed needs of the region (hugely exaggerated) had been fulfilled. Valencia, which had be compensated from the failed Ebro project, claimed the right to receive water from other basins. Castillala Mancha (where the headwaters of the Tagus-Segura project are located) put a date on the termination of the *Trasvase* to the Southeast. Faced by such a landslide of demands for exclusive decision making by the regions, the Spanish state has taken many of the regional laws to the Constitutional Court, while the regions are engaged in a bitter struggle to gain the power to decide about the final destination of the flows.

4.6 CONCLUSIONS

Water politics, water culture and water engineering have all played a central role in shaping the Spanish landscape and society. The contemporary water geography and ecology of this country is the product of centuries of socio-ecological interaction. Neither the history of the country nor its present geographical layout can be understood without taking into account the radical transformation of the water landscapes. Spain holds the world record in the largest geographical area *per capita* occupied by man-made reservoirs. In terms of governance, a key factor in the development of Spanish water policy has been the early establishment of a solid policy community consisting of irrigators, hydroelectric power companies and engineering firms. These interested parties have been able, during the twentieth century, to take over the administrative and regulatory bodies, blocking the adaptation of the water policy to evolving social demands.

An important lesson that can be drawn from the Spanish case is that long-term water policy almost exclusively oriented towards the continuous increase of supplies is not free of contradictions. Overcoming of scarcity as a short or mid-term goal, based on increased river impoundment, may bring only temporary benefits while shortages and conflicts are pushed into the future.

Overestimation of resources, demands for meeting short-term management goals, and appeasement of pressure groups inevitably result in social or environmental crises in the medium- or longer term. In Spain, this practice has resulted in the over-allocation of existing resources and pushing into the future of politically difficult decisions to limit demand and use. New plans must deal with the absence of new supply augmentation alternatives; challenges to implement in-stream flow requirements, and decisions involving inevitable trade-offs.

With respect to spatial water allocation decisions, the Spanish case, with a strong tradition of river basin-based water planning and management, exemplifies the conflicts inherent to interregional transfer decisions. As we have noticed, this is particularly true when these transfers move water from one river basin to another, and even more so when the water transfers affect regions with different economic, social, political and administrative structures. Conflict is directly related to a collective sense of inequity in the allocation decisions and, increasingly, to the defence of environmental and patrimonial values in the donor basins.

Citizen and stakeholder involvement in transparent and well informed participation processes is a key element of adaptive water management. However, the Spanish case shows that the demand for increased social participation is difficult to implement. It requires changes in mentalities and power structures. It also implies the acknowledgement of the inevitable convergence of national interests with management perspectives that acknowledge and defend regional territorial interests. These are still unsolved problems that are sure to shape Spanish water politics and practice in the years to come.

REFERENCES

Arroyo Ilera F. and Camarero Bullón C. (1989). *Proyectos ilustrados de navegación fluvial* (18th Century River navigation projects). In *Los paisajes del agua* (Water Lanscapes), Universitat de València and Universidad de Alicante, Valencia, 347–369.

Fernández Ordoñez J. A.(coord.). (1986). *Catalogo de treinta Canales Españoles anteriores a 1900* (A Catalog of thirty Spanish Channels built before 1900). Comisión de Estudios Históricos de Obras Públicas y Urbanismo, Colegio de Ingenieros de Caminos, Canales y Puertos, Madrid.

Garrido A. and Llamas R. (2009). Water Policy in Spain. London: Taylor and Francis.

Gil Olcina A. (2002). *Perduración de los Planes Hidráulicos en España* (Persistence of Water Planning in Spain), http://habitat.aq.upm.es/boletin/n27/aagil.html#fntext-1 (Last accessed 10th December 2011).

Gil Olcina A. and Rico Amorós A. (2008). *Políticas del Agua II. Mejora y ampliación de los riegos de Levante* (Water Policies II. Expansion and improvements of irrigation in Eastern Spain). ESAMUR, Murcia.

Gómez Mendoza A., Sudrià C. and Pueyo J. (2007). *Electra y el estado: la intervención pública en la industria eléctrica bajo el Franquismo* (Electra and the State: government intervention in the electricity industry under Franco's regime). Thompson–Civitas, Madrid.

Gómez Mendoza J. (1992). *Regeneracionismo y regadíos* (Regenerationism and irrigation projects). In: *Hitos históricos de los regadíos españoles* (Historical milestones of Spanish irrigation), A. Gil Olcina and A. Morales Gil (eds.), Ministerio de Agricultura, Pesca y Alimentación. Madrid, pp. 231–262.

López Gómez A. (1989). *La presa y el canal del Guadarrama al Guadalquivir y al Océano. Una utopía fallida del siglo XVIII* (The dam and the canal from the Guadarrama River to the Guadalquivir River and the Ocean. A failed eighteenth century utopia), Bol. Real Acad. de la Historia, T. CLXXXVI, Cuaderno II, pp. 221–262.

López Gómez A. (1992). *Presas y canales de riego en los siglos XVI y XVII* (Dams and irrigation channels in the sixteenth and seventeenth centuries). In: *Hitos históricos de los regadíos españoles* (Historical milestones of Spanish irrigation), A. Gil Olcina and A. Morales Gil (eds.), Ministerio de Agricultura, Pesca y Alimentación, Madrid, pp. 91–142.

López Ontiveros A. (1998). *El regadío, salvación de la patria y fuente de felicidad, según los congresos nacionales de riegos (1913–1934)* (Irrigation as salvation of the country and source of social happiness, according to the National Irrigation Congresses (1913–1934)), *Demófilo. Revista de Cultura Tradicional de Andalucía*, **27**, 7–64.

Molina Ibáñez M. and Montiel Molina C. (2004). *Desarrollo y repercusiones del parque hidroeléctrico en los regímenes fluviales* (Development and impact of hydroelectricity in flow regimes). In: *Alteración de los regímenes fluviales peninsulares* (Alteration of Iberian flow regimes), A. Gil Olcina (coord) (ed.), FundaciónCajaMurcia, Murcia, pp. 177–195.

del Moral Ituarte L. (2010). The hydraulic paradigm and production of a new geography in Spain: pnorigins and historical evolution between the Sixteenth and Twentieth Centuries. In: A History of Water. Series II Volume 2: Rivers and Society: From Early Civilizations to Modern Times, T. Tvedt and R. Coopey (eds.), I.B. Tauris, London, pp. 440–462.

Nadal E. (1993). *Introducción al Análisis de la Planificación Hidrológica* (Introduction to the Analysis of Water Planning). Ministerio de Obras Públicas y Transportes, Madrid.

Ortega Cantero N. (1984). *Las propuestas hidráulicas del reformismo republicano: del fomento del regadío a la articulación del plan nacional de obras hidráulicas* (Hydraulic proposals of Republican reformism: from the promotion of irrigation to the national Water plan). Agricultura y Sociedad, **32**, 109–152.

Smith N. A. F. (1970). The Heritage of Spanish Dams. Spanish Committe of Large Dams, Madrid.

Sauri D. and del Moral L. (2001). Recent developments in Spanish water policy. Alternatives and conflicts at the end of the hydraulic age. *Geoforum*, **32**, 351–362.

Swyngedouw E. (1999). Modernity and hybridity: regeneracionismo, the production of nature and the Spanish waterscape, 1890–1930. *Annals of the Association of American Geographers*, **89**(3), 443–465.

Tvedt T. (2009). Water systems' perspective on history. In: A History of Water. Series II, Volume 2: Rivers and Society: From Early Civilizations to Modern Times, T. Tvedt, R. Coopey and I. B. Tauris (eds.), London, pp. 3–28.

Chapter 5

Wither water: reflections on Colorado's water development and the pursuit of sustainable growth in the arid west

Troy Lepper

> *"Humanity has the ability to make development sustainable to ensure that it meets the needs of the present without compromising the ability of future generations to meet their own needs. The concept of sustainable development does imply limits—not absolute limits but limitations imposed by the present state of technology and social organization on environmental resources and by the ability of the biosphere to absorb the effects of human activities. But technology and social organization can be both managed and improved to make way for a new era of economic growth."*
>
> —Brundtland Report, 1983

5.1 INTRODUCTION

As we move firmly into the 21st century it is becoming clear, especially in arid environments, that competition for scarce water resources will be the primary driver of social development. Water has always been a constraining factor on social development in the western part of North America. Even the most drought conditioned societies have fallen prey to its desiccating effects. From approximately A.D. 750 – A.D. 1300 the Anasazi Indians were a proud and vast nation in this region. It is now understood that the Anasazi were extremely adept at controlling the flow of the limited water they had available in their settlement areas. The Anasazi dug retention ponds to capture runoff, and they performed routine maintenance on these small reservoirs to mitigate silt build up. The Anasazi use of retention ponds to capture water for future use allowed the settlement at Mesa Verde to thrive and grow at a time when most of the southwest region was experiencing population declines, but the use of retention ponds still did not allow the Anasazi to escape from the limits of an annual average of 45.7 centimeters (18 inches) of precipitation. In the end, from around A.D. 1135 – A.D. 1300 (Wright, 2004:7) prolonged and repeated drought finally broke the tribes resolve to stay at Mesa Verde driving many of its inhabitants northwest into what is now New Mexico and Colorado.

There is much to be learned from this example of social development and failure, especially today. Population growth and technological advancement go hand in glove when it comes to the provision of water, and even though it may seem we have sustainably moved past the carrying capacity of the regional arid environments we inhabit in the western United States, that sustainable reality can be turned on its head with the arrival of a severe and prolonged drought.

5.2 THE CASE OF COLORADO

In 2000, Colorado, and the rest of the western United States began experiencing the impacts of a weather shift that resulted in drought conditions for almost everyone in the region. The occurrence of drought is common in this region, and was by no means the most severe drought on record. In retrospect, the drought that crushed the Anasazi was more severe and prolonged than what we experienced starting in 2000, but by the end of 2002 many municipalities in Colorado were getting nervous. It does not take long in an arid environment to see the impacts of drought when reservoir levels start to drop, crops and lawns start to die off and rivers dry up. Not only does water dry up, but so do the local and regional economies that rely on that water. The Anasazi and other ancient societies that settled in this area most likely experienced these same problems, only on a smaller scale. In the year 2011 the idea of small scale development is just that, an ideal. This point is made crystal clear when we look at the development of the State of Colorado in reference to its water resources, and the demands put on those resources as social development took hold, and finally the technological solutions implemented when those precious drops of water could no longer sustain growth and development.

5.2.1 From conflict to codified law: the development of the doctrine of prior appropriations

When people from the eastern part of the United States first migrated to Colorado they were immediately confronted with the harshness of an environment that only produced 45.7 centimeters of precipitation a year. This migrating population was more conditioned to humid environments where dry land farming practices were sufficient to bring in enough food to feed your family and have a little excess to sell at market. It immediately became clear that the only way to survive under these conditions of water scarcity was to irrigate their fields with ditches off the main stem of any river they could find. This temporary technological solution allowed new communities like Union Colony and Fort Collins Agricultural Colony to take hold and grow, but drought quickly brought these two communities to the brink of violence when Fort Collins Agricultural Colony moved upstream from Union Colony and began diverting Cache la Poudre River water when very little river water was available. Fort Collins Agricultural Colony ended up diverting nearly the entire river, leaving Union Colony with no water to grow their crops. As Mark Twain famously stated, "whiskey is for drinkin', water is for fighting over", and this was certainly the case with Union Colony and Fort Collins Agricultural Colony. This potential disaster was averted when unexpected rains arrived and cooled tempers down, but the potential for other violent water conflicts loomed over the newly developing State of Colorado, therefore a solution was legislated into the state's constitution called the Colorado Doctrine. This doctrine defined four core principles of Colorado water law:

(1) "All surface and groundwater in Colorado was a public resource for beneficial use by public agencies, private persons and entities;
(2) A water right was a right to use a portion of the public's water resources;
(3) Water rights owners could build facilities on the land of others to divert, extract, or move water from a stream or aquifer to its place of use; and
(4) Water rights owners could use streams and aquifers for the transportation and storage of water." (CFWE 2003:5)

These core principles laid the legal foundation for what was referred to as the Doctrine of Prior Appropriation which can be loosely interpreted as "first in time, first in right, as long as you can put the water to a beneficial use." It was this doctrine that allowed the water of the state to move around the landscape whether that water moved from the west slope to the east slope of the continental divide[1], or it moved from one basin to another. Originally the State of Colorado only recognized three beneficial uses (agricultural, municipal and industrial), but as the state grew, and its economic and social demands on water diversified, so too did the allowable beneficial uses. For example, "In 1973, the State Legislature recognized the need to correlate the activities of mankind with some reasonable preservation of the natural environment. To accomplish this, it created the Instream Flow and Natural Lake Level Program within the state agency, the Colorado Water Conservation Board." (CFWE, 2003:7) Unlike states governed by riparian water rights that tie the water to the land that it was adjacent to, Colorado allowed the water to move from one place to another based on a system of water rights ownership and priority. This meant water rights that were established early were superior to those established later, which reiterated the driving principle of "first in time, first in right", and in the State of Colorado, as was the case in many states in the arid west, agricultural water rights usually had priority.

As you can see in Figure 5.1, Colorado is projecting its water use by sector where agricultural water use continues to dominate the water landscape by consuming approximately 82% of the State's usable water, Municipal and Industrial (M&I) water use accounting for 15% of the State's usable water and Self-Supplied Industrial (SSI) accounts for just 3% of the water. Even though agricultural water use continues to consume the majority of Colorado's renewable and nonrenewable water resources, the State is projecting agriculture's impacts on water to decrease as population growth, accompanied by municipal and industrial growth, draw Colorado's scarce water resources to higher and more efficient economic demands. In the end, water flows to where the money is and with the decline of agriculture in the U.S. in general, and Colorado in particular, we can expect this trend to continue.

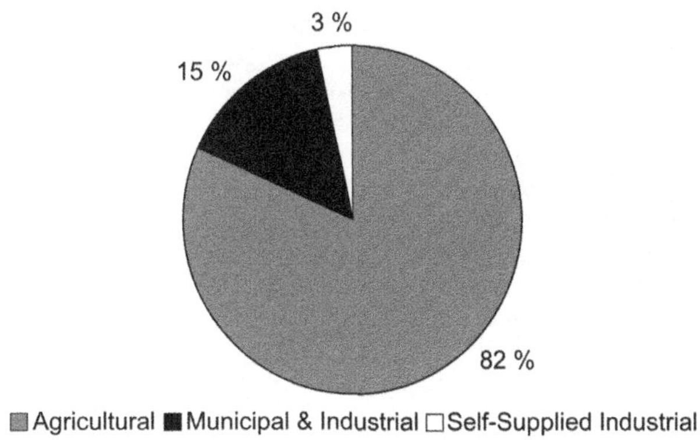

Figure 5.1 Projected Water Use by Sector by the Year 2050 SWSI Fact Sheet: 2.

[1]It is important to understand that 80% of Colorado's water and 20% of its population is located on the western slope of the continental divide, leaving 20% of the water and 80% of the population on the eastern slope of the continental divide. The majority of the state's population lives along the Front Range of the Rocky Mountains in the cities of Pueblo, Colorado Springs, Denver, Boulder and Fort Collins.

5.2.2 Interstate compacts and the doctrine of prior appropriations

The management of Colorado's great rivers has been influenced by many challenges, but one challenge has been with the state since its inception, and will most likely present the most serious challenge to Colorado, simply because Colorado cannot control how other states will challenge Colorado's claim to its headwaters in the coming years. It is interesting to observe in Figure 5.2 the amount of water that Colorado must allow to flow to downstream states. "The state of Colorado has rights and obligations at the headwaters as the mother of many of the great rivers. Nine interstate compacts, two Supreme Court equitable decrees and two Memoranda of Understanding govern how much water Colorado is entitled to use and consume within its boundaries." (CFWE, 2010:3) These compacts and legal agreements have forced states in the arid western U.S. to become strange bedfellows. Each compact is different, and must be examined as separate agreements. One compact may have a set amount of water that must be delivered to downstream states, like we find with the Colorado River Compact. The Colorado River Compact allocates 7.5 million acre feet of Colorado River water to be delivered to 4 Upper and 3 Lower Colorado River Basin states. This may seem fair on the surface, but a couple of things must be considered before you pass judgment. First, when the Colorado Compact was negotiated by Delph Carpenter in 1922, the region was experiencing a very wet period that resulted in more water being in the river when it was divided up than the river, on average, has seen since. Another problem was the stipulation that "the Upper Basin states may not cause the flow of the Colorado River at Lee Ferry, Arizona, to be depleted below an aggregate of 75 million acre feet for any period of 10 consecutive years." (CFWE, 2010:1) This fixed amount of water that must be delivered to Lower Basin states could prove to be problematic, especially in a region that is experiencing climate changes that could result in a 20 percent reduction in the flows of the Colorado River in the coming future.

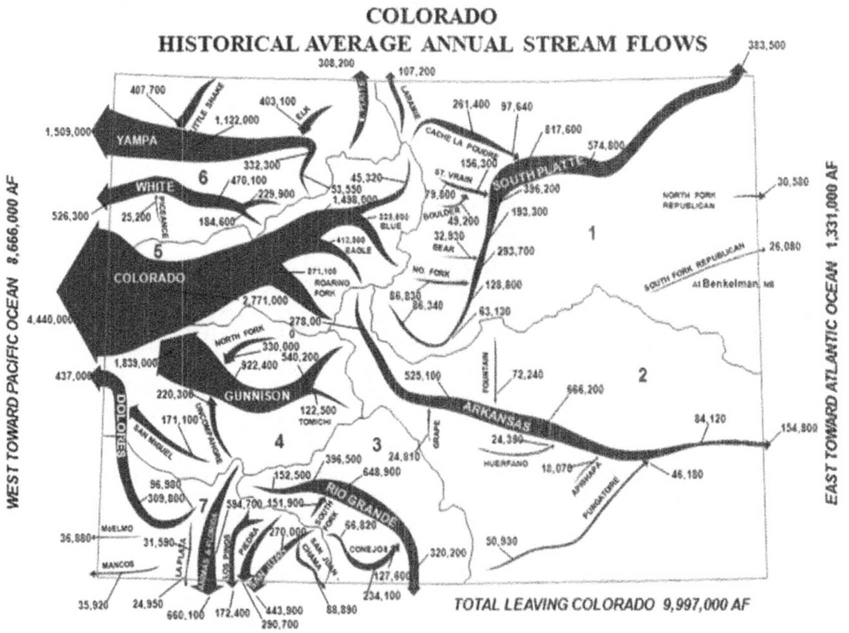

Figure 5.2 Colorado Historic Average Annual Stream Flows (prepared by Hydrographic Branch of the Office of the State Engineer Colorado Division of Water Resources with 2011 revisions) http://water.state.co.us/SurfaceWater/SWRights/WaterDiagrams/Pages/SnakeDiagram.aspx.

The Colorado River Compact is very different than the South Platte River Compact negotiated in 1923. This compact apportions the South Platte between Colorado and Nebraska, giving Colorado full use of the flows of the South Platte between October and May, but during the irrigation season, "Colorado must deliver 120 cubic feet per second to Nebraska at the Julesburg gauge. If the flow is less than 120 cubic feet per second, Colorado must curtail junior diversions." (CFWE, 2010:1) As you can see, there is no prescribed compact formula to follow. This process must be negotiated with a wide variety of stakeholders representing a wide variety of interests. There are no quick fixes in state, national and international agreements, but if the negotiated agreements are done thoughtfully with an eye to future generations, and power is distributed equitably to stakeholders throughout the negotiations, then compromises made on the negotiation table can actually lead to interstate and international agreements that are respectable, enforceable and binding in a way that stands the test of time.

In many respects the Doctrine of Prior Appropriation, which was originally hammered out in the context of water conflicts in the state of Colorado was used to appropriate the flows of Colorado's rivers within, as well as between, states in the developing western U.S. This should come as no surprise, since the problems associated with water scarcity were essentially the same for developing communities as they were for developing states, only the size and scale were different.

5.2.3 Climate change and reoccurring drought cycles and their impact on water management

Another specter complicating water management in the State of Colorado is that of climate change. Up to this point, we have talked about Colorado's experience with drought in reference to regional shifts in weather patterns that are not necessarily tied to changes in the Earth's climate, even though Colorado's regional weather patterns may be exacerbated by the onset of climate change in the entire system. "Climate change challenges the traditional assumption that past hydrological experience provides a good guide to future conditions" (Bates *et al.* 2008). With the onset of climate change, we can expect more uncertainty and unpredictability in the State's available water resources. "An underlying assumption of traditional hydrologic frequency analysis is that climate, and hence the frequency of hydrologic events, is stationary, or unchanging over time. Anthropogenic climate change (i.e. processes derived from human activities) and better understanding of decadal and multi-decadal climate variability present a challenge to the validity of this assumption." (Kiang *et al.* 2010:1)

Historically water managers have treated climate as stationary, therefore making it easier to model the management of the State's water resources with some level of predictability. But the onset of climate change is now calling into question the practice of treating climate as stationary forcing water managers into a new paradigm known as nonstationarity. (Hirsch, 2010; Stakhiv, 2010; Webb & White, 2010) This recognition of nonstationarity has forced modelers to recalibrate their projections. "With respect to assumptions of stationarity, the engineering community of practice should be concerned about changing the methodologies of assessment to better incorporate the uncertainty introduced by the lack of stationarity and to develop acceptable methods to project future conditions." (Webb & White, 2010) These new projections are proving problematic for the State of Colorado for at least 3 reasons:

(1) Unlike most states whose statewide climate is relatively uniform, Colorado's climate varies tremendously from place to place.

(2) Colorado's topography presents challenges for modelers. Up to this point, due to limitations of the models, modelers have treated the Rocky Mountains as a broad hump, which fails to take in consideration the variability in micro-climates from the high mountain tops to the low mountain valleys.
(3) Lastly, our recent models are predicting an expansion northward of deserts in the southwest with storm tracks following suit, but variability in different models leaves water managers uncertain about how these changes are actually going to play out. (Udall, 2008:20)

In the end, when it comes to climate change, Colorado is going to experience uncomfortable changes that will present significant challenges to water managers. Statewide temperatures will continue to rise. Over the past 30 years Colorado's average temperatures have risen approximately 1 degree Celsius (2 degrees Fahrenheit), and they are expected to continue to rise another 1.25 degrees Celsius (2.5 degrees Fahrenheit) by 2025. This temperature rise will produce milder winters and a predicted reduction in snowpack. (Udall, 2008:26) We have already seen changes in the arrival of snowmelt as spring streamflows. Spring runoff is arriving approximately 14 days earlier (Udall, 2008), creating challenges for the allocation and distribution of Colorado's scarce water resources that remain in line with Colorado's Doctrine of Prior Appropriation.

5.2.4 Water management and the complexities of urban growth in arid climates

Compounding the problems of weather and climate related issues in the arid western U.S. are massive increases in municipal and industrial growth. As we see in Figure 5.3, in Colorado, the extensive water works necessary for facilitating that growth are found on both sides of the continental divide, and are comprised of a series of interlocking rivers, ditches, reservoirs and tunnels. California blazed this trail nearly a century ago, and as the rest of the states in the west slowly developed, they in turn were confronted with the need to develop their water resources in order to keep pace with growth. As we have seen throughout history, this treadmill of water production represents a slippery slope. When societies use technology to solve water shortage problems in arid climates, they allow for continued population growth, which eventually brings that society right back to the water stressed place it started. If we are to believe the Statewide Water Supply Initiative findings, then we can expect Colorado's population to nearly double by 2050 (SWSI, 2010). Most of this growth will continue to take place along the Front Range in the Arkansas, South Platte and Denver Metro River Basins. Even though most of the population growth will happen on the eastern slope of the Rocky Mountains, the highest rates of growth will actually take place on the western side of the divide. "Growth rates on the west slope will be as high as 240 percent, whereas on the Front Range the growth rate will be approximately 70 percent."(SWSI, 2010:3) This growth will only place more stress on the available renewable water resources needed to meet existing, as well as future demands. "Colorado, in turn, will need between 600,000 and 1 million acre-feet of additional municipal and industrial water by 2050." (SWSI, 2010:3)

As you can see in Figure 5.4, the M&I gap will not be distributed equally between Colorado's river basins. Some river basins like the Gunnison can meet most of their projected M&I demands with existing and identified projects, whereas other basins like the Denver Metro Basin will have a difficult time closing the gap between existing supply and future demand. But what is most striking about Figure 5.4 is that no river basin in Colorado is projected to completely close the M&I gap.

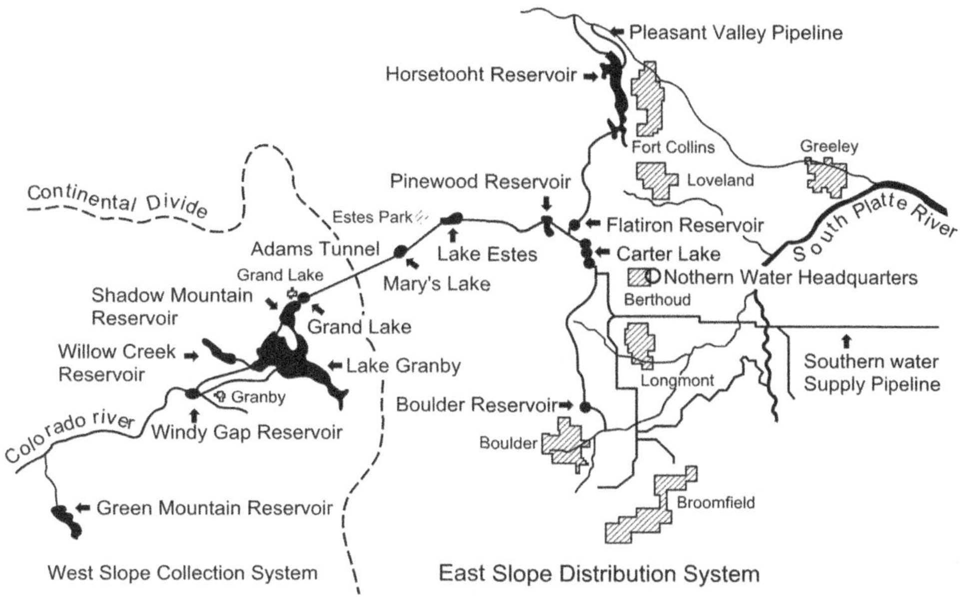

Figure 5.3 Map of the Northern Colorado Water Conservancy District with Projects Servicing Growth along the Front Range of the Rocky Mountains in Colorado (NCWCD website 2012) http://www.northernwater.org/default.aspx.

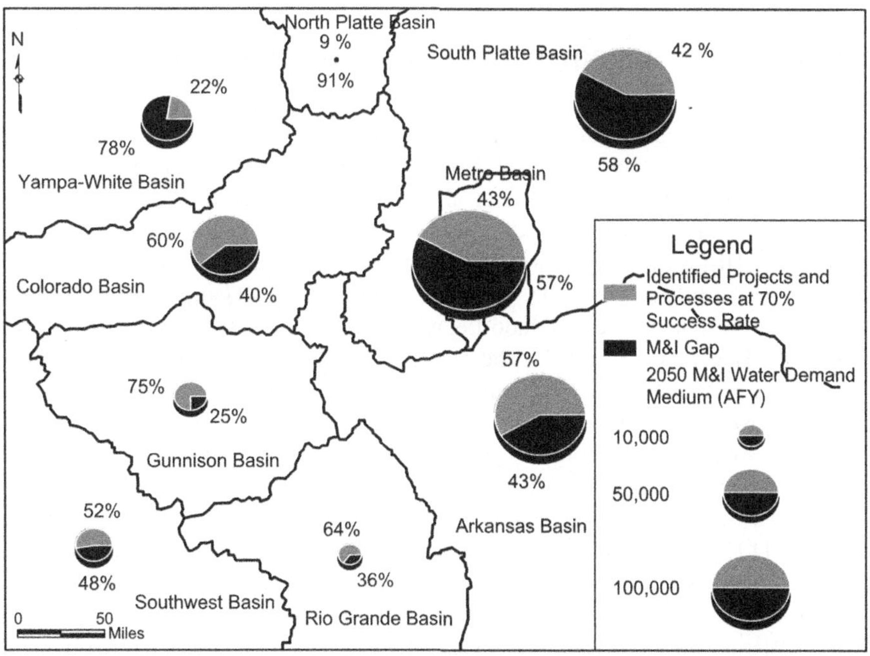

Figure 5.4 Colorado's 2050 Municipal and Industrial Gaps per River Basin (SWSI, 2010:5–30).

This is where things get interesting, especially in a state that legally sanctions the movement of water around the landscape through formal and informal water markets. It has been a long recognized fact that the South Platte River Basin has been running one of the largest water markets in the world. Water has been bought, sold and leased in the South Platte River Basin in accordance with Colorado water law since the arrival of miners searching for gold in the Rocky Mountains. These markets have literally ebbed and flowed in reference to how well society used technology to solve the age old problem of water scarcity. Technology has been used to draw water out of the rivers for immediate use, as well as storing that water in surface and subsurface reservoirs for later use. Technology has also been used to transfer water from one basin to another using canals, tunnels, reservoirs and pumps. Inter-basin water transfers diverting water from rivers on the western slope of the Rocky Mountains to the eastern slope allowed Front Range communities to continue to grow, even though they were running up against the natural carrying capacity of their local environment.

But when it comes to population growth and water development, enough is never enough, and by the middle part of the 20th century, Colorado's Front Range water users were once again water short. By this time Coloradoan's had come to rely on technological solutions to their water shortage problems, and as they did in the past, they once again looked to technology to save the day and this time it was the electric pump that pushed continued social development through the environments carrying capacity. The electric pump allowed water users the option of drawing water directly out of the ground, which allowed a much more efficient application of water than was previously possible through irrigation canals. As you can imagine, problems began to surface quickly. Senior surface water rights owners immediately noticed river levels dropping when pumps came on in the late 1950s and by the late 1960s Colorado was forced to integrate and regulate surface and groundwater under the Doctrine of Prior Appropriation with the passage of the 1969 Colorado Ground Water Management Act. Once this new wrinkle in water management was ironed out, population growth soared.

5.2.5 Colorado water management in the 21st century: planning and assessment

But as we have seen time and time again in arid climates, drought is the great equalizer. During the 1980s and 1990s Colorado weather systems produced above average precipitation amounts that filled reservoirs and continued to push the limits of growth, but all that changed when drought arrived on Colorado's doorstep in early 2000, and by 2005 Colorado's water surplus had turned into a deficit and water panic was starting to set in. Historically when water conflicts sprung up during times of water stress, water users called on lawyers to litigate the battle, but at the urging of Russ George, Director of the Colorado Department of Natural Resources, the Colorado Legislature blazed forward with a different path with the passage of Colorado House Bill 05-1177: The Colorado Water for the 21st Century Act. This time Colorado was going to tackle the water problem head on, and they were going to rely on a collaborative intra- and inter-basin roundtable process that brought the wide range of stakeholders to the negotiation table to hammer out solutions to Colorado's present and future water shortages. In 2003, it was also recognized that if the State of Colorado was going to push forward with regional and statewide water planning, then it was imperative that a statewide assessment of Colorado's water portfolio was conducted. HB 05-1177 was launched in conjunction with the Colorado Statewide Water Supply Initiative to take a comprehensive look at Colorado's water resources, and how those water resources can be used to sustain growth and development for generations to come.

The Colorado Water for the 21st Century Act created intra-basin roundtables for resolving the conflicts over water shortages in the State as a whole using a basin by basin strategy. Each basin assigned local water stakeholders to roundtables, and those stakeholders began to assess the problems their basins were confronted with due to projected water shortages, but they did not stop there. Once the problems were identified, they then moved forward with plans to prioritize their solutions. It is important to remember, when it comes to water shortages there are no easy solutions to the problems each individual basin faces. This fact is complicated by the longstanding conflicts between water user organizations that have dominated basin politics for years. Once each river basin roundtable began to take ownership of the complex problems associated with projected population growth and weather and climate changes, the state could then bring representatives from each basin roundtable to begin the more arduous process of hammering out a statewide water plan that would limit conflict between basins as Colorado continues to grow and develop under conditions of water stress and scarcity.

5.2.6 Colorado water management in the 21st century: implementation

Colorado has now moved this process from the assessment and planning stage to the implementation phase. The state has identified 4 potential strategies for meeting the projected M&I gap, none of which are new to water users in the State of Colorado:

(1) Agricultural Transfers
(2) Development of New Supply Options
(3) Conservation
(4) Identified Projects and Processes

Agricultural water transfers hold potential for reducing the M&I gap, but they also present problems that are not easily addressed. Return flow requirements, water quality problems and the inability to move water efficiently upstream make much of the water that flows to agriculture near impossible for any other use. Return flow requirements are particularly problematic, because the conditions placed on an agricultural water right usually require the non-consumptive portion of the water right to be delivered to its historical point of delivery around its approximate historical time of use.

New supplies are also considered a potential solution to closing the M&I gap, but most of these options require construction or enlargement of reservoirs, which in 2011 can be a problem. Not only are the NEPA requirements for constructing large-scale dam and reservoir projects more stringent than they were in the U. S. golden age of dam building, but the public perception of building large-scale water storage projects has also shifted. During the Great Depression the federal government supported the building of projects like the Hoover Dam in order to harness the nation's water ways for development, but it was also using these projects to literally jump start the nations failing economy by putting Americans to work on building public works projects. There was very little thought given to the species that would become endangered or go extinct because of the engineering of our nation's waterways. This change in public perception is exemplified in the case of the potential construction of the Northern Integrated Supply Plan (NISP) for Northern Colorado[2]. Northern Colorado's potential for development is directly connected to existing water supplies, and the potential for acquiring more water in the future. Most water managers in the Northern Colorado Water Conservancy District understand the need for more storage, and much of the support for NISP actually comes out of the basin roundtable process, but when you bring the general public into the

[2]Maybe the most important instance of negative public perception preventing the construction of a large-scale dam project in the State of Colorado, that would have provided enough new water for development along the Front Range well into the 21st century, is the case of Two Forks. In the 1990s the powerful Denver Metropolitan Governments, represented by the Denver Water Board, were defeated by a local and national coalition of environmental groups set against seeing another natural treasure, in this case Waterton Canyon on the South Platte River, sacrificed for unbridled growth.

debate, things that would normally have passed fifty years ago with little resistance can now be thrown off track through social protests and legal challenges. NISP seemed to be on the fast track for passage and development, but shortly after the Environmental Impact Statement was published a vocal local group of residents formed a group called Save the Poudre and slowed the process down to a crawl. As of November 2011, it was still uncertain whether NISP would ever make it off the drawing table, but without NISP, or other storage projects like NISP, Colorado will have a difficult time closing the M&I gap.

Conservation is always an option for lessening the M&I gap. "The Colorado Water Conservation Board defines water conservation as those measures and programs that provide for measurable and verifiable permanent water savings." (CWCB, 2010:7–7) Encouraging water users in arid environments to use less water is always necessary, but all water conservation is not good conservation. As was stated earlier, agricultural water conservation has limited potential due to the conditions placed on agricultural water rights and the non-consumptive return flows associated with those rights. Urban water conservation is a different story. Cities can implement a number of strategies to decrease water consumption, and the conservation savings can make a real impact in reducing the M&I gap, but conservation does not solve the problem of population growth, it only pushes it off into the future. The real question for cities in Colorado is what do they do with their conserved water, and when/if population once again outpaces water availability, can the State rely on conservation to make up the difference.

The final potential option is Identified Projects and Processes (IPP). "IPPs are defined as projects and processes local water providers are counting on to meet future water supply needs." (CWCB SWSI Fact Sheet, 2010:4) This category includes many of the options we have already discussed, and lumps them into an overall strategy for prioritizing and implementing a number of different projects simultaneously in order to meet the varied state's projected water management needs in the present and future. "The Interbasin Compact Committee and Colorado Water Conservation Board found that implementing the IPPs is critical to minimizing the water supply gap, but IPPs should be implemented in a way that balances the state's responsibilities to protect and restore Colorado's natural resources." (CWCB, 2010: 7–7) These projects include agricultural water transfers, reuse of fully consumable water, growth into existing supplies, regional in-basin projects, new transbasin projects, firming in-basin water rights, and firming transbasin water rights.

5.3 CONCLUSION

So how is Colorado doing in reference to the Brundtland's call for sustainable development? Has Colorado followed through with its ability "to make development sustainable to ensure that it meets the needs of the present without compromising the ability of future generations to meet their own needs?" (Brundtland, 1983) If the Statewide Water Supply Initiative findings are a good indication of the state of our existing and future water supplies seen in reference to existing and projected demand, then the answer to that question is no. Every basin will be water short in the year 2050, which in turn will stretch existing water supplies even more, and without new technological or social solutions to these water management problems growth will have to slow. Complicating these matters is the specter of drought and climate change. Severe drought has the potential to bring megacities in the arid west to their knees since until recently those growing cities have not been planning sufficiently for the quick and lasting impacts of drought. One thing is certain, social development cannot exist without water, and sprawling megalopolises like those we find along the Front Range of the Rocky Mountains in Colorado have a new found respect for that reality. Colorado water managers are now recalibrating their models, policies and practices to address the uncertainty inherent in climate change and population growth, but only time will tell whether our sprawling societies can withstand the devastating effects of prolonged drought, in order to sustain development for our generation without compromising the future of those who come after us.

REFERENCES

Bates B. C., Kundzewicz Z. W., Wu S. and Palutikof J. P. (eds.). (2008). Climate Change and Water: Technical Paper of the Intergovernmental Panel on Climate Change. IPCC Secretariat, Geneva, Switzerland.

Brown K. (2003). Citizens Guide to Colorado Water Law. 3rd edn., Colorado Foundation for Water Education, Denver, Colorado.

Cantwell R. (2010). Citizens Guide to Colorado's Interstate Compacts. Colorado Foundation for Water Education, Denver, Colorado.

Hirsch R. (2010). A Perspective on Nonstationarity and Water Management. Presented at the Workshop on Nonstationarity, Hydrologic Frequency Analysis, and Water Management in Boulder, Colorado. Colorado Water Institute Information Series No. 109, Fort Collins, Colorado, pp. 515.

Kiang J., Olsen R. and Waskom R. (2010). Executive Summary Presented at the Workshop on Nonstationarity, Hydrologic Frequency Analysis, and Water Management in Boulder, Colorado. Colorado Water Institute Information Series No. 109, Fort Collins, Colorado, pp. 1–3.

Stakhiv E. (2010). Practical Approaches to Water Management under Climate Change Uncertainty. Presented at the Workshop on Nonstationarity, Hydrologic Frequency Analysis, and Water Management in Boulder, Colorado. Colorado Water Institute Information Series No. 109, Fort Collins, Colorado, pp. 20–28.

Statewide Water Supply Initiative (2010). Colorado's Water Supply Future. Colorado Water Conservation Board Denver, Colorado. http://cwcb.state.co.us/water-management/water-supply-planning/Documents/SWSI2010/SWSI2010.pdf (accessed 16 August 2012).

Statewide Water Supply Initiative (2010). CWCB SWSI 2010 Mission Statement, Key Findings, and Recommendations. Colorado Water Conservation Board, Denver, Colorado, p. 4. http://cwcb.state.co.us/water-management/water-supply–planning/Documents/SWSI2010/SWSI2010FactSheet.pdf (accessed 16 August 2012).

Udall B. (2008). Global Climate Models: What They Do and Do Not Show. Citizens Guide to Colorado Climate Change. Colorado Foundation for Water Education, Denver, Colorado.

United Nations Report of the Brundtland Commission (1983). Process of preparation of the environmental perspective to the year 2000 and Beyond. General Assembly Resolution, 38/161, p. 24, 19 December 1983.

Webb J. and White K. (2010). Nonstationarity in Water Management: USACE Perspective. Presented at the Workshop on Nonstationarity, Hydrologic Frequency Analysis, and Water Management in Boulder, Colorado. Colorado Water Institute Information Series No. 109, Fort Collins, Colorado, pp. 16–19.

Wright K. and Wright R. 2004. Ancient Puebloan Reservoirs of Mesa Verde Citizens Guide to Colorado's Water Heritage. Colorado Foundation for Water Education, Denver, Colorado.

Chapter 6

Water recycling and sustainability in Salisbury, South Australia

Carol Fort

Sitting in South Australia watching the cooling evening rain in summer 2011, it is difficult to remember that only two years ago south-eastern Australia was nearing the end of the worst drought in its weather records, which had already lasted over a decade. At that time, no-one knew when the El Niño weather system would release Australia from its grip, and South Australia's normal rainfall-based water supplies, including the Murray-Darling Basin, were perilously low.

During the drought, a significant project in the City of Salisbury, a northern suburb of South Australia's capital city, Adelaide, showed the world how systematic stormwater recycling could augment urban water supplies and help preserve the dwindling flow in drought-stricken river basins. Popularly known as Waterproofing Northern Adelaide (although that title officially describes only part of the project), the scheme directly addressed the issue of environmental sustainability in a living context of municipal government decision-making and local provision of potentially potable water in a political system where water supply is a public-private monopoly business. This chapter examines the historical record of the project, asking questions about how governments might make sustainable environmental policies and what factors support successful implementation of those policies.

The research shows that successful water-sustainability policy is possible, but durable and resilient sustainability policy is more likely to come from situations where there are well-defined component problems with a local focus, a wide variety of personnel with a wide variety of expertise-perspectives (not just the typical experts), a flexible approach to spending money – probably enhanced when the money comes from multiple sources – and a way of capturing whatever moral high ground is available.

6.1 INTRODUCTION

6.1.1 Thinking about sustainability

The idea of sustainability has been evolving for many decades in the sense of the "maximum safe yield" concept associated with activities such as forestry and fishing (Tortajada, 2003) and was picked up in early UN statements on development. In its contemporary form, the idea of sustainability is relatively new, hitting the natural and social sciences together in 1987 when the Report of the World Commission on Environment and Development called for sustainable development, or "development that meets the needs of the present without compromising the ability of future generations to meet their own needs" (WCED, 1987). Since then, the world has recognised sustainable development as its most pressing dilemma, acknowledging and addressing it in the Millennium Statement, and making it the focus of the United Nations Decade of Education for Sustainable Development (UNESCO 2005). This Decade for Change, running from 2005 to 2014, has the overarching goal of changing world attitudes to sustainability. Changed attitudes are essential precursors to sustainable life on earth because sustainable development will only result from open debate and compromise of stakeholders' competing interests, but in a world already exhausted by the virulence of climate change debates with their attacks on science and divisiveness in public policy, thinking about sustainability is complicated by several difficulties. Although we have the well-known definitions, the word still has no agreed meaning. Worse, the meanings it does have are so overwhelming that they lie outside the reach of normal policy-making whatever sort of government a society has and however well-meaning it might be.

The reason for its overwhelming nature is that sustainability is an abstract idea, but is also an absolute – like truth or justice. This leaves thinkers with three problems that interfere in discussions of sustainability, especially across national or cultural boundaries. First, each of us conceptualises sustainability in slightly different ways even when we speak the same language. Our attitudes alter the meaning we impute to it, and those attitudes tend to be marked by mind-sets prepared by our personal histories and experiences and by our current circumstances. Sustainability is a value and it is very hard for us as social beings to separate our own social, economic or political identity from our value judgments. So, the meanings we give to sustainability are ideologically driven (Soderbaum, 2000; Tortajada, 2003). Second, growing from its ideological basis, the concept of sustainability has a morality about it – if you can claim to be sustainable, somehow you are "better" and can revile your less sustainable neighbours. The moral flavour of the sustainability concept allows discussion to degenerate into name-calling and simple disputes which concentrate on whether a thing is "good" or "bad" according to pre-judged criteria rather than on close examination of the case under discussion. This is a particularly tenacious problem because most people get their sustainability ideas and information from popular media, which prefers to streamline complex ideas into dichotomy. The third, and most significant difficulty, is that sustainability feels as though it has an absolute nature. A practice seems to be either sustainable or unsustainable. The lack of middle ground frightens people, especially policy-makers. Consequently, they feel, you can't have a little bit of sustainability. You can't revisit the definition of sustainability (development that meets the needs of the present without compromising the ability of future generations to meet their own needs) and amend it to read: development that compromises only some of the future generations or development that compromises only sectors of future generations. You can't do that because the result wouldn't be sustainability, it would be, in fact, the situation we have now where unsustainable practices impact unevenly on different socio-economic or political groups. Even though this last factor is clearly not a practical definition, which would recognise some practices as more sustainable than others, the difficulties remain. Even where there is stakeholder agreement on the meaning, designing and implementing operational schemes is fraught with conflict. So, for example, the world agrees that water problems of shortage and quality must be addressed but successful programmes remain rare.

6.1.2 Water in South Australia

In Adelaide, as in most Australian cities, the amount of rain that falls on the city is roughly equal to the amount of water the city uses. This highlights the challenge thrown down by UNESCO, that "water crisis is primarily one of governance", because, the IHP documents tell us, "institutions lack the capacity to overcome conflicting approaches in the use and allocation of water" (IHP, 2008). Water sustainability is, therefore, primarily a system of government that prioritises water sustainability over competing water demands. This is clear in Australia. Overall, Australia has a relatively small population in a large landmass, which means that Australia itself is not short of water, a reality that has led Australians, even those living in the dryer states, such as South Australia, to ignorance of water's value and has obscured the fact that parts of Australia are very dry. In South Australia, only 3.3% of the land area receives average rainfall of over 500 mm a year, while 87% receives less than 250 mm (Atlas, SA, 2000). Even in the temperate zone where Adelaide is situated, El Niño weather patterns cause severe droughts from time to time, some lasting many years. Not only do El Niño conditions diminish rainfall in the Adelaide region, but more significantly, they bring drought into the Murray-Darling Basin, which regularly augments Adelaide's rainfall-fed drinking water supply.

Settler South Australia has always suffered from shortage of water and from poor quality of the water that was available. From the very beginning in 1836, South Australian governments have worked hard to convince migrants, investors and businesses that the government could give them water security. Here we are in 2012 and the problem remains the same but worse. Not only have we established a society with an unsustainable thirst for water, but we have started to feel the impact of shifting climate and, intensifying our risk, we are sitting on the same ticking bomb as many parts of the world: unprecedented rates of population rise and urbanisation.

6.1.3 The project in context

This chapter looks at a local government storm-water recycling scheme that aims to make that local government area (the City of Salisbury, a northern suburb of Adelaide) more self-reliant in water supply: in other words, to implement a policy of sustainability. The project aims to capture a significant part of its stormwater and other runoff and recycle it through a system of detention in engineered wetlands, storage in natural aquifers and recovery for later use, saving up to 30 gigalitres a year of piped potable water. Water supply in the state of South Australia is a state government responsibility and the water services (supply and infrastructure) are currently provided to Adelaide by a partnership of the government and a commercial monopoly owned by a suite of global water companies, SA Water. Responsibility for stormwater, the subject of this chapter, lies outside SA Water's purview, remaining firmly in the hands of local municipal governments.

Salisbury is a huge municipal government area covering 161 square kilometres, with landscapes including hill faces, seashores, mangrove forests, military bases, residential areas new and old, shopping centres, old and new industry, salt-fields, market gardening, and much else. It is largely situated on the northern Adelaide Plain, an almost completely flat flood plain between the scarp of the hills to the east and the placid, gulf coast at its western extremity, with some residential development mounting the adjacent hills face. In the manner typical of arid lands, swollen creeks rush from the hills after periods of torrential rain and spread themselves over the flats where drainage is slow and incomplete. Before the beginning of twentieth-century urbanisation, the area covered by Salisbury was marshy wetland, but after white settlement began in 1838 gradual hardening of increasingly urbanised surfaces increased the speed of runoff with consequent decreasing quality of surface and ground water on the plain and at the coastal margins.

Since beginning the extensive harvesting and recycling project, Salisbury now has a complex of over 50 artificially constructed wetlands that slow down and clean storm water and attract wildlife (see Figure 6.2), pumping stations that force harvested water down into the underlying aquifer (Figure 6.1) and recovery mechanisms that make that water available to the local water market. Most important, Salisbury has realistic plans to increase the harvest and yield and has provided consulting expertise to other governments both in Australia and overseas. The research question is a simple why? Why has this scheme grown and integrated? Why has it shown such success?

Figure 6.1 Managed Aquifer Recharge injection pump at Greenfields Wetlands, Salisbury.
Source: Photograph May 2012, Carol Fort.

Figure 6.2 Constructed wetlands slow the flow of stormwater and give multiple benefits. Not only is the water cleaned before injection into the aquifer, but the ponds also provide pleasing landscapes for humans, habitats for many animals, and welcome resting places for popular visitors such as this pelican.
Source: Photograph May 2012, Carol Fort.

6.1.4 Historical case study method

To analyse this question, I am using the case study method. Case Study is a very common methodology for historians – a very good method to use when the boundaries between the phenomenon being examined and its context are not clear. It allows for multiple sources of evidence and many different kinds of evidence to be brought to bear on the analysis. So, for example, this case examines the Salisbury recycling experience through scientific evidence, landscape evidence, government records, newspaper reports and personal interviews. The case study method, in analysing and interpreting the particular example in its own context, allows research to focus on the processes at work, particularly strategic decision making and the impact of factors such as path dependence.

6.1.5 Water recycling through aquifer recharge, storage and recovery

The stormwater management programme in Salisbury is technically known as Managed Aquifer Recharge (MAR) with components of Aquifer Storage and Recovery (ASR), often referred to as ASTR, or Aquifer Storage, Transfer and Recovery. Essentially, this means a programme of gathering stormwater or other recyclable runoff into ponds and then injecting it underground into a suitable natural ground water storage called an aquifer (see Figure 6.1). In Salisbury's project, stormwater is not simply used to recharge the aquifer, but it is stored there for recovery and use later. Most important, the water is not simply recovered through the injection pipe, as in most MAR cases, but has separate input and extraction bores, forcing the water to move and guaranteeing it will remain underground for a predictable time. While underground, natural processes attenuate nutrients, degrade trace chemicals and kill pathogens (CSIRO, nd). A significant part of the scheme, the Salisbury "ASTR stormwater to potable water project" is a demonstration project managed by a partnership of the City of Salisbury, the South Australian Government, the Australian Government, United Water, SA Water and CSIRO as part of a European Union project, "Reclaim Water". It demonstrated in June 2007 that after a year underground the recovered water met drinking standards (Dillon *et al.* 2009).

MAR, with or without associated storage and recovery, has been used all over the world but attitudes towards it and the relative success of various projects varies from place to place. This international context makes this a good time to examine MAR projects and makes Salisbury a fruitful case study.

6.2 RESEARCH FINDINGS

6.2.1 Sustainability by design?

The major finding is entirely predictable: Salisbury's sustainability policy is the result of its experience with successful practice. Recently erected advertising signs on the road leading into the City of Salisbury past the Greenfields Wetlands (see Figures 6.2 and 6.3) read: "Our City Saw the Future of Water and Harnessed It". But this is not how the project began. In the strict sense, the scheme was not policy-driven, or even sustainability-driven. As Brooks and Holtz point out in their soft-path analytics, soft paths, or routes that break the path dependence typical of water supply, begin when planners start by defining a sustainable future and then define the appropriate policies and programmes that will bring them into being (Brooks & Holtz, 2009:88). The findings of this research show that Salisbury did not start by pursuing a sustainability goal: no-one ever thought: "let's create a lovely series of wetlands and use the water to recharge our dying aquifer and then recover some of it to convince industry to stay in our area and maintain our employment levels". Salisbury dealt with one problem at a time, often placing contradictory solutions side-by-side. Thus the research finds no simple, smoothly integrated answer. No particular aspect of the project guarantees the success of all, but several components seem to be very important in the total picture. They include a raft of environmental problems the Council was bound to be addressing anyway, creative, innovation-focused staff, very real rises in the cost of customary water (increase has been five-fold since 2009), a moral boost from conserving water in drought years, significant industry and science partners, large tracts of land available for experimenting with engineering wetlands in a landscape that was naturally marsh and wetland in pre-development times, and suitable aquifers. In other words, the complexity and extent of Salisbury's programme grew entirely out of local circumstances. That said, the research does show that as the scheme expanded in size, it expanded in vision. While the project did not begin by planning sustainability, and sustainability was certainly not mentioned in early strategic plans, the historical record shows that the scheme's potential for supporting sustainability planning was recognised and was actively pursued, not only in Salisbury but elsewhere (see Section 3 below).

Figure 6.3 Publicity from the City of Salisbury, showing the multiple uses of the Greenfield Wetlands – including nature trails, a café, toilets and picnic tables – and also indicating the new emphasis on sustainability.

6.2.2 "The paddocks": An experiment

During the 1960s, Salisbury's urbanisation moved off the flats and onto the low block-hill face to the City's east, interfering with drainage out of the higher territory. This encouraged flooding and pooling on waste land and Council decided to convert some of that land. Landscaped and planted, in 1975 it became the first constructed wetland, known as the Paddocks, and council learned that it encouraged wildlife and made a nice recreational area. More significantly, Council learned that in slowing down floodwater runoff, the water could be made to drop the heavy metals it picked up off the road into the ponds and reed-beds, ensuring that the discharge into watercourses was healthier (City of Salisbury, nd).

6.2.3 Pipes, surface water and groundwater

In 1987, a consortium headed by the South Australian government began to look for a site for a futuristic technology park and residential development. They decided to build what they called a Multifunction Polis on degraded land, some of which was within Salisbury's Council area. This meant there was a conglomeration of Federal, State and Local Government interest in the area. When the idea fell through in 1998, mainly because of the lack of foreign investment, lots of the local money was still available with no demand that the investment produce commercial profits. The land is now occupied by Mawson Lakes, a suburb of Salisbury, which is a brand-new residential development which connects the much-vaunted "third pipe" to all houses. The third pipe carries non-potable water, which comprises a combination of recycled sewerage and recovered MAR stormwater, for toilet flushing and outside use (Mawson Lakes, nd).

At the time, many of the engineers brought in to the Multifunction Polis site were gas engineers from the north of the state (Kaufman, 2009). This meant that they were men with experience of pipelines and aquifers, able to think fluidly about moving water from one end of the council district to another without flinching. These men brought together two seemingly disconnected problems. First, a long-standing problem: generations of market gardeners on the plains to the north-west of Salisbury had drawn off the aquifer to the extent it was pulling in sea water – the death of the industry was just around the corner at a time when Adelaide's galloping population rise was demanding ever-increasing vegetable production. Second, they saw sewerage. Adelaide's major sewerage treatment plant is at Bolivar near the coast in the Salisbury council district. Public demands that the plant must stop discharging into the sea, met the market gardeners' need for another source of water. In 1994, the Council and the Department of Mines drilled the first ASR bore into the aquifer to relieve it with refreshed stormwater. Later, cleansed Bolivar water was added to the supply. The Virginia Pipeline Scheme was commissioned in 1999, according to SA Water, the first and largest of its type in Australia (SA Water, nd).

6.2.4 Environmental quality

Focus on the Multifunction Polis site drew attention to water quality and drainage in a more general way. Salisbury is a coastal council hosting some of the most valuable mangrove forests in the state. It also houses an extensive salt/chemical industry with evaporation pans on part of the tidal margin, which needs to prevent land waste-water from entering its operation. Thus, stormwater was diverted around the plant in the 1930s, intensifying its impact on surrounding inlets. At about the same time Salisbury was experimenting with its early constructed wetlands, conservationists, representatives of the fishing industry and developers at adjacent Port Adelaide combined to demand Salisbury stop sending its stormwater down the drainage system to places like the Barker Inlet where the mangrove forests and sea-grass meadows are the major nursery for South Australia's marine fisheries. In 1993, the Council began to construct wetlands at Greenfields, interfering with the drainage pattern leading from Salisbury's roads and lowering the nutrient and pollutant load of stormwater that did reach the Inlet (Salisbury City Council, nd).

6.2.5 Commercial interests

In 1997, a major industrial water user, the largest wool-processing plant in Australia GH Michel, threatened to leave Salisbury, taking hundreds of jobs with it. Why? The company wanted to move somewhere where water would be cheaper – it was using one billion litres (1 gigalitre) of mains water every year and producing a similar volume of dirty water. Salisbury Council, using land rented from an airport in the Council area, were able to create a stormwater diversion system with accompanying ASR that could replace the mains water used by the company. Further, the recycled water was so much lower in salt content than the mains water that it reduced Michel's costs even further (Hamnet, 2003).

6.2.6 The moral high-ground

The most significant element underpinning the project's successful implementation was the drought itself. After the experiences of success described above, drought conditions carried the ideas forward. The prolonged drought in a milieu of rising climate change awareness forced South Australians to examine their water use. Extreme, government-mandated water-use restrictions were applied: garden watering and car washing were forbidden, governments handed out flow-limiting shower heads, municipal councils replanted public gardens with drought-resistant native plants and redesigned their watering regimes, and South Australians daily discussed their water recycling and re-use strategies. Radio talk-back, television news and newspapers kept the issue of water-shortage alive in everybody's minds while the federal government subsidised house-hold rainwater tanks, supported local council re-use projects, categorised appliances for water-efficiency, sponsored water-awareness-raising in schools across the nation, and, after over 200 years of white settlement, finally began to develop a management plan for the Murray-Darling Basin. Government slogans such as "water wise" and "water for good" (SA Government, 2009) added a frisson of morality and intelligence to the daily pragmatics of catching household grey water in buckets to pour on suffering lawns and rose-bushes. The drought gave sustainability the moral high ground it might not have otherwise gained.

Partly, the rectitude was imported from elsewhere. The drought's impact on the Murray-Darling Basin was so intense it achieved political significance. Upstream over-allocation to commercial users combined with increased demand from downstream Adelaide forced the federal government to formulate an integrated approach to the basin's waters. This was heralded by massive advertising campaigns in which federal and state government statements about "caring" for and "protecting" the Murray were incorporated into local thinking and supported South Australians' new awareness of urban water. Salisbury Council used the ideas to scaffold its recycling scheme, including "providing a reliable, alternative water supply for the future" in its own advertise (City of Salisbury, 2010). This "supply for the future" idea touches on the local link between sustainability and climate change. The Salisbury district is extremely exposed to projected rises in sea-level. While rate-payers might have resisted paying for drainage work to enhance mitigation of flooding that *might* happen in the future, in the main they happily paid for drainage work that fitted into the general belief that recycling is a morally good thing. Recently, the council incorporated this into a wider sustainability framework, adding the phrase "Salisbury, Sustaining Our Environment" to its publicity and renaming its wetlands information centre the "Salisbury Sustainability Precinct" (City of Salisbury, nd). Figure 6.3 is part of a recent brochure publicising the Salisbury Sustainability Precinct and the Greenfields Wetlands. It clearly shows the change in Salisbury's approach as it positions the wetlands in the wider "sustainability" framework. The brochure also clearly indicates the multiple benefits of wetlands: not only water refreshing installations, but havens for wildlife and enhanced landscapes for humans too. This wider environmental identity, enhanced by images of frogs and dragonflies on advertising material, connects the wetlands to global environmental advertising. The Greenfields Wetlands, featured in Figures 6.2 and 6.3, is the most visible locally. Situated abutting a junction of two major highways, it is well known for its flocks of water birds, including the large, dramatic and easily seen pelicans and black swans. Although the bird numbers were less in the years following the drought – with water again available in the birds' customary habitats – the idea of saving the birds enhanced the general rectitude of saving the water.

6.3 SOME CONCLUDING THOUGHTS

So. Here we have a series of problems associated with water (flooding, expense, pollution of ground water and surface waterways) combined with a series of unrelated factors specific to the locality: failed government urbanisation projects, failing industries (including wool processors, fisheries and market gardens), unanticipated money, a strange assortment of people who were capable of seeing problems through new eyes, and a drought that made a virtue out of water recycling.

Analysis of the Salisbury responses to these problems shows that durable and resilient sustainability policy is possible but is more *likely* to come from situations where there are well-defined component problems, easily identified connexions to the locality, a wide variety of personnel with a wide variety of expertise-perspectives (not just the typical experts), a flexible approach to spending money – probably enhanced when the money comes from multiple sources – and a way of capturing whatever moral high ground is available. Looking at a comparable project helps to highlight the Salisbury findings.

Similar to Salisbury, the City of Onkaparinga is a large, peri-urban, coastal suburb of Adelaide with very mixed land use, but, as Salisbury is situated in the far northern suburbs of Adelaide, Onkaparinga is situated to the South. A recent planning document shows attention to familiar matters.

Water Proofing the South is a localised integrated water resource management strategy that was initiated by the City. It aims to provide alternative water sources such as reclaimed water and stormwater to replace the use of mains water and groundwater, and is currently being delivered in two stages (City of Onkaparinga, 2011).

The two stages include elements common with Salisbury's experience: public-private partnership, effluent reuse, "capture, storage, treatment and reuse of stormwater through an integrated system of Managed Aquifer Recovery", innovative thinking in Council personnel, extensive pipe transfers and about $22 million of government money.

Although Waterproofing North and South required significant injection of external funds, both benefitted from the fact that, in the South Australian government structure, stormwater and most other forms of runoff are the responsibility of local councils. The localised nature of these recycling programmes makes them particularly suited to local governance and particularly responsive to ongoing decision-making related to feedback and changing circumstances. This local responsiveness to changing circumstances in the wider context is evident in the way the Salisbury Council began to associate its water recycling projects with the wider "sustainability" context, for example. Problems such as flooding, poor water quality and agricultural degradation of groundwater are problems local councils were expected to solve in some way so thinking about them creatively and innovatively was, in some senses, expected by existing policy. Both councils had vast public spaces – parks, gardens, sporting grounds and school properties – that they were watering at ever-increasing expense to council rate-payers. When Salisbury began recycling, producing irrigation water was not particularly cheap. Recently, rapid water-price rises associated with state-wide infrastructure and growing pressure from climate change have made the financial costs more competitive but, in addition, the environmental benefits in the locality are priceless to residents.

The Onkaparinga strategy shows that despite the local circumstances that nurtured the development of Waterproofing Northern Adelaide, elements of the programme have proved durable and transferrable. "Water Wise Adelaide" is a discussion paper produced by the statutory body created to manage the northern project and the government money that funded it. In 2008 the paper recommended that the state government support an Adelaide-wide application of Salisbury's principles, stating:

> *South Australia's prosperity, possibly more than any other State, is critically linked to the provision of a sustainable water supply. That prosperity and future growth is now at risk with little time left to resolve the threat.*
>
> *The driest state in the driest continent now faces the very real prospect of not having enough water to sustain the city's basic needs within 25 years. The River Murray is no longer a dependable supply of drinking water, climate change is resulting in unpredictable rainfalls and Adelaide catchments are under pressure from urban growth.*
>
> *This proposal [of expanding a Salisbury-like programme across Adelaide] is based on the recent success of Waterproofing Northern Adelaide and the knowledge gained in establishing Salisbury's network of wetlands. This experience has shown that if all parties work together, the community can be assured that their future water needs will be met (Waterproofing Northern Adelaide, 2008).*

Thus a programme that began from scattered small-scale local interventions is now a world leader, not only in scientific practise but in sustainability also. While it might not be true that Salisbury "saw the future of water and harnessed it" as proclaimed in advertising material, it is certainly true that the City realised the potential of its early schemes and recognised the "sustainability" concept as a powerful driver for their amalgamation.

REFERENCES

Brooks D. and Holtz S. (2009). In the beginning: soft energy paths. In: Making the Most of the Water We Have: The Soft Path Approach to Water Management, D. Brooks, O. Brandes and S. Gurman (eds), Earthscan, Sterling, VA, pp. 35–48.

City of Onkaparinga Climate Change Responses in Southern Fringe of Adelaide Case Study 18/11/11, (consulted 24 May 2012).

City of Salisbury (2008). Sustaining our Environment http://www.salisbury.sa.gov.au/Our_City/Environment/Sustainability_and_Climate_Change, (consulted 24 May 2012).

City of Salisbury (2010). The Salisbury Wetlands postcard, personal collection.

City of Salisbury nd http://www.salisbury.sa.gov.au/Our_City/Environment/Wetlands_and_Water, (consulted 24 May 2012).

City of Salisbury Paddocks Wetlands http://www.salisbury.sa.gov.au/Our_City/Environment/Water/Wetlands/Wetlands_Locations/Paddocks, (consulted 24 May 2012).

CSIRO (2010). Managed Aquifer Recharge: Frequently Asked Questions http://www.csiro.au/ Organisation-Structure/Flagships/Water-for-a-Healthy-Country-Flagship/~/media/CSIROau/Flagships/Water%20for%20a%20Healthy%20Country%20Flagship/ManagedAquiferRecharge_WfHC_pdf%20Standard.pdf, (consulted 24 May 2012).

Dillon P., Pavelic P., Page D., Beringen H. and Ward J. (2009). Managed Aquifer Recharge: An Introduction. Commonwealth Government of Australia, National Water Commission, Canberra.

Hamnet S. (2003). Adelaide as a Sustainable City: Water Management in the Broader Context of Sustainability, paper to the Sustainable Urban Services Shanghai Seminar, Pacific Economic Cooperation Council, Sustainable Cities Task Force, China National Committee for Pacific Economic Cooperation.

International Hydrological Programme, Division of Water Sciences IHP-VII Water Dependencies: Systems under Stress and Societal Responses 2008–2013, UNESCO.

Kauffman C. Executive officer of Waterproofing Northern Adelaide, personal communication, August 2009.

Mawson Lakes Sustainability nd, http://www.mawsonlakes.com.au/llweb/mawsonlakes/main.nsf/all/ml_aboutus_su, (consulted 24 May 2012).

PPK Consultants *et al.* (1992). Gillman/Dry Creek urban development proposal: draft impact statement. Department of Environment and Planning, Adelaide.

SA Water, historical timeline http://www.sawater.com.au/nr/rdonlyres/b34d3058-b0bd-4134-88c7-f5bd51e209ff/0/sawater_150y_book.pdf, (consulted 24 May 2012).

South Australian Government (2000). Atlas of South Australia http://www.atlas.sa.gov.au/go/resources/atlas-of-south-australia-1986/environment-resources/water-supply, (consulted 24 May 2012).

South Australian Government (2009). Water For Good and 'Waterwise Communities' http://www.waterforgood.sa.gov.au/using-water/waterwise-communities/, (consulted 24 May 2012).

Stevenson L. and Barbara L. (1990). The Multifunction Polis: A Catalyst for Change in the 21st Century. University of Wollongong Academic Services Division, Wollongong.

Tortajada C. (2003). Rethinking development paradigms for the water sector. In: Rethinking Water Management: Innovative Approaches to Contemporary Issues, C. Figuères, C. Tortajada and J. Rockström (eds), Earthscan Publications, London, pp. 8–23.

UNESCO Education for Sustainable Development http://www.unesco.org/new/en/education/themes/leading-the-international-agenda/education-for-sustainable-development/, (consulted 24 May 2012).

Waterproofing Northern Adelaide Regional Subsidiary (2008). Water Wise Adelaide Executive summary of submission to South Australian Water Security Council, November 2008. http://www.civictrust.net.au/WaterWiseExecutiveSummary.pdf, (consulted 24 May 2012).

World Commission on Environment and Development (1987). Our Common Future report 42/187 http://www.un.org/documents/ga/res/42/ares42-187.htm (consulted 24 May 2012).

Chapter 7

Acid mine water drainage in South Africa: policy, economic issues and public concerns

Johann WN Tempelhoff and Frank Winde[1]

7.1 INTRODUCTION

Over a period of more than 160 years, South Africa has built up a reputation as one of the world's premier mining countries. The modern industry began with copper mining operations in Namaqualand in the north-western Cape in the early 1850s (Solomon, 2011). Towards the end of the 1860s, diamond mining started at Kimberley in the Northern Cape. In the 1880s mining activity was extended to the Witwatersrand; this frenzy of early profit making paved the way for the founding of Johannesburg, the City of Gold – home to the financial region where the lion's share of the world's gold was mined for the greater part of the twentieth century (Buranelli, 1979). Besides diamonds and gold, an array of other precious and non-precious minerals – including uranium (Winde & Sandham, 2004) – have been added to the impressive list of South Africa's mining output. Of particular importance for the country's export trade over the years have been: coal mining, which began in the 1870s (De Jager, 1976); platinum mining since 1924 (Cousins, 1976); and mining of iron ore since the 1950s (Hammerbeck *et al.*, 1976). Collectively the minerals made a substantial contribution towards making the South African economy one of the strongest on the African continent.

In recent times the narrative of what used to be a heroic discourse of mining industrial development has undergone a marked change. There is now a tendency in civil society to be reflective and cynical vis-á-vis the mining sector (Turton, 2012). At the heart of the matter is the issue of water. South Africa is one of the driest countries in the world and its available water resources are limited (Research Channel Africa, 2011).

There are growing concerns that acid mine drainage (AMD), a toxic polluted water that emanates from worked-out mining operations, poses a major threat to South Africa's water resources. And this is not only of growing importance in South Africa; the mining sector in all parts of the world is grappling with the problem. Indeed, as early as the 1980s the United States' Environmental Protection Agency identified

[1]School of Basic Sciences North-West University (Vaal), South Africa. Thanks to Dr. Bridget Theron for language editing the manuscript.

AMD as a problem that could potentially have a negative impact on the environment, in much the same idiom as other global issues such as climate change and the earth's disintegrating ozone layer (Manders *et al.* 2009).

Public concern about the potential contamination of South Africa's existing water supplies is well-founded. The country has an average annual rainfall of 450 mm – well below the world average of 860 mm/annum – and relies to a considerable extent on surface water for most of its supplies. As a result of an increasing demand and the vagaries of anticipated global climate change, scientists are predicting critical water shortages when the currently available supply peaks by 2025 (Archer, 2010; Turton, 2012).

The environmental awareness of South Africans has been increasing since the 1990s. Now public opinion is sympathetic to discourses on the potential of tourism against mining activity to bolster the country's future economic development. Articulate and well-informed members of civil society, many of whom have joined forces with local and international non-governmental organisations (NGOs), have of late taken a firm stand against proposed mining activities. One example is the concerted campaign in opposition to plans for fracture mining operations for shale gas ("fracking") in South Africa's dry Karoo region. The major concern is that local water resources in an already fragile arid environment will come under threat (Turton, 2012). The debate has been informed by what has been happening in Gauteng Province's Witwatersrand region where AMD and potentially dangerous heavy metals have been decanting from the premises of worked-out gold mining sites. The toxic water then finds its way into streams and lakes before meandering into the large rivers of the South African interior.

But what is AMD? Essentially it is caused by pyrite (FeS2), also known as "fool's gold". When pyrite is exposed to air and water it forms sulphuric acid, iron oxides and hydroxides that are typically found in the mineral goethite.[2] The profound influence of the chemical process is evident in water when the pH level drops to about 4. The oxidation process is accelerated by bacteria, of which *Thiobacillus ferrooxidans* is a major contributor. The acid produced by the pyrite reacts with the rock or residue deposit and then forms salts, mobilising heavy metals that may be in the rock or in the residue. Water contaminated by AMD is toxic to fauna and flora (Akcil & Koldas, 2006; McCarthy, 2011; Wells *et al.* 2009).

The potential threat of AMD in South Africa is significant. The mining industry currently consumes 6 percent of South Africa's available water supplies in scattered regions covering about 7 percent of the country's surface area (Wells *et al.* 2009). The areas most directly affected are the exhausted gold and coal mining regions situated in the Witwatersrand region of Gauteng Province, as well as Mpumalanga, North West, KwaZulu-Natal and the Free State Province. In some areas where mining operations are still under way, the problem appears to be under control, but only if mine water is pumped, at a significant energy cost, to specially prepared storage facilities. However, most water, up to the present, has been discharged into nearby streams with only small quantities used for tailings disposal.

7.2 AMD AND THE GOLDFIELDS OF THE WITWATERSRAND

On the Witwatersrand, most of the underground gold mining operations have come to an end after more than 125 years. The closure of mines has had a marked influence on the natural water supplies. When deep-level mining operations began in 1889, water had to be pumped from vast natural dolomitic subterranean chambers in the western central and eastern parts of the Witwatersrand, to make it possible for the miners to extract the gold-bearing rock underground. By the 1930s, as mining operations had to be continued ever deeper, it was necessary to mine below dolomite. Consequently, the François

[2]The mineral goethite in acid mine water is responsible for the red/orange colour of the sediment in a stream or a water body affected by AMD: $Fe(OH)_3$ in chemistry terms.

cementation process was developed to seal shafts that had been sunk through water bearing karst (Winde & Stoch, 2010b; Stoch & Winde, 2010). The potential of toxic contamination of water also remained as an increasing range of chemicals such as cyanide were used, along with the even more poisonous substance, mercury, in the gold extraction process (Pogue, 2006).

In the mining and industrial environment of the early twentieth century, the water issue was addressed swiftly when in 1903 the Rand Water Board (currently Rand Water) was formed as a regional bulk potable water supply utility in a collaborative initiative between the mining industry, local authorities and the central government (Tempelhoff, 2001). Since then, there has been a consistent supply of clean drinking water in the sprawling industrial environment that has evolved as the economic hub of South Africa. Planners and managers of infrastructure, supported by the government, made sure that near-pristine quality water supplies were pumped from distant localities. At first water was pumped from the Vaal River, about 70 km to the south of Johannesburg. Then followed a series of comprehensive dam construction projects (Tempelhoff, 2006) and water transfer schemes to Gauteng from KwaZulu-Natal, the Free State and eventually as far afield as the highlands of Lesotho (Haarhoff & Tempelhoff, 2007). Currently Rand Water is by far the largest water utility in South Africa providing bulk water supplies of a high quality to the entire Gauteng and parts of the North West, Mpumalanga and Free State Province.

Mine water was also managed well. Some of the water was used to transport tailings onto dams. It was recovered via so-called penstocks and routed into 'return water dams' for being re-used. None of the water was treated before discharge into adjacent streams.

For the greater part of the twentieth century, the government nurtured the gold mining industry. The sector generated substantial taxable profits and the state enjoyed a reliable source of income. Furthermore, mining created opportunities for secondary industries, trade, commerce and finance to flourish. Importantly, the mining sector also created job opportunities, so when the cost of gold mining production rose in the 1950s, the government actively supported schemes to provide financial assistance. By the 1960s the state started subsidising the costly process of pumping water from the mine shafts (Thorius, 2004). The measure was recommended to government by an interdepartmental commission (1956–1960) that had been working on the implication of dewatering underground chambers for the purposes of mining activities (Stoch & Winde, 2010).

Given the international political pressures that were brought to bear on the South African government as a result of the ruling National Party's discriminatory racial policies of apartheid, the alliance between the state and the mining industry was of crucial importance. Politically, South Africa was, as a result of its strategic position as a supplier of minerals, especially uranium – a by-product of gold mining on the West Rand (Maziza, 1993) – tolerated by some of the leading Western powers. Thus, the state had some good reasons for consistently supporting the mining sector and also providing the necessary subsidies to keep pumping mine water from underground to prevent acid mine drainage from affecting mining output. This continued well after the 1970s, when some of the worked-out gold mines on the Witwatersrand began closing down (Figure 7.1).

The infrastructure *status quo* of pumping water proved to be fragile over the longer term. Once the mining operations ceased, water pumping also came to an end. The cavernous voids created by the extensive deep-level mining are now gradually filling up. The water flowing into the old mines is largely rainwater, which contains dissolved oxygen that turns acidic as it goes through mine waste deposits on surface and underground the mine passages. It is also enriched by heavy metals. Once the voids fill up they begin to decant at the lowest-lying points of the old mines (McCarthy, 2011). This process has been taking place on the Gauteng West Rand near Krugersdorp and Randfontein, affecting the Wonderfonteinspruit and Tweelopiespruit, respectively. In the next couple of years AMD will also decant from other areas of the Witwatersrand.

Figure 7.1 A storage dam on the West Rand filled with acid mine drainage notable for high levels of uranium. *Source*: Photograph JWN Tempelhoff February 2012.

7.2.1 The political factor of nationalisation

Politically, the current state of affairs has been shaped by a historical dynamic that had its origins in the mid-twentieth century. As early as the 1950s, the African nationalist resistance against the white minority government was actively articulating its opposition to a mining industry that was wholly owned and managed by strong capitalist interests. In its Freedom Charter, adopted at Kliptown, near Johannesburg in 1956, the African National Congress (ANC) and its alliance partners subscribed to the principle of state ownership of the country's mining sector for the benefit of all the country's people. However, in the aftermath of the fall of the Berlin Wall in 1989 and the collapse of the East-West political divide, the ANC leadership has consistently accepted the principles of a free market economy in a world that has moved strongly towards the principles of liberal democracy and the recognition of capitalism. In the process, the principles of the Freedom Charter of "nationalising" the country's mines have been toned down (Klein, 2012). In the negotiations that preceded the transition to a multiracial democracy in 1994, there was general acceptance of the fact that the state would accept the responsibility of acting as the custodian of the country's mineral resources on behalf of the people without "nationalising" the mines. Subsequent legislation approved by parliament in consultation with all stakeholder sectors also put an end to the private ownership of mineral rights. Mining activity, it was agreed, would in future be subject to tacit approval in the form of certification and licensing arrangements between the government and mining companies.

7.2.2 Relevant legislation

The *Minerals and Petroleum Resources Development Act* (MPRDA), No. 28 of 2002, is an important piece of legislation controlling mining activity. Since its implementation in 2004, the MPRDA has been the key to enforcing the guidelines for environmental protection, but it also stipulates the role of the state and the ownership of the country's minerals. The MPRDA repealed the *Minerals Act*, No. 50 of 1991 and regulates prospecting and mining as well as petroleum exploration and mining in South Africa (Wells *et al.* 2009).

The MPRDA's objectives include:

- Recognising the internationally accepted right of the state to exercise sovereignty over all mineral and petroleum resources in the area under its jurisdiction;
- Accepting that the state is the custodian of the country's minerals and petroleum resources on behalf of the people;
- Promoting equitable access to the nation's minerals and petroleum resources to all the people in the country;
- Working towards meaningfully expanding and promoting opportunities for historically disadvantaged persons;
- Promoting the growth and development of the mineral and petroleum resources of the country;
- Promoting employment and advancing the socioeconomic welfare of all South Africans; and
- Ensuring that there is security of tenure for prospecting, exploration, mining and production operations (Wells *et al.* 2009).

In terms of the Constitution (1996), the MPRDA should also ensure that the minerals and petroleum resources are developed in an orderly and ecologically sound manner. The legislation also has to ensure that the holders of mining and production rights contribute to the socio-economic development of the areas in which they operate (Wells *et al.* 2009).

One implication of the new dispensation of 1994 was that the state no longer had the responsibility to summarily provide support to the mining companies by, for example, subsidising water pumping from the mine shafts. In many parts of the Witwatersrand, the costly pumping activities were therefore halted and in 2002, when the first acid mine drainage started decanting into the Tweelopiespruit (Stoch *et al.* 2011) on the West Rand, there was a public outcry. A major concern was the high level of radiation coming from large quantities of the heavy metal uranium that was present in the water. In view of the fact that the Wonderfonteinspruit is one of the waterways in the Vaal River catchment, and that it flows through farming and urban areas, there was considerable consternation.

7.2.3 Contending with the Witwatersrand AMD problem

Numerous research projects and innovative plans have been mooted to address the problem. In 2009 the Western Utilities Corporation (WUC), established with the expertise of specialists at South Africa's Council for Scientific and Industrial Research (CSIR), tabled plans for a mine water treatment plant that would produce 60Mℓ of industrial grade processed AMD water for consumers in Gauteng. The objective was to recover a range of by-products such as sulphur and metals, leaving behind little or no residual waste (Manders *et al.*). However, practical issues prevented the project from being fully implemented.

Dealing with acid mine drainage has been fraught with difficulty for both the government and the mining industry. Many of the original companies that had been at the helm of the mining sector for decades have sold their mines to new companies, most of which were created and flourished within the framework of the government's policy of black economic empowerment (BEE). The government's objective was to promote the interests of the upcoming entrepreneurial sector of previously disadvantaged South Africans. As the existing mine management vacated their positions and these were filled by new managers and operators, crucial human resource skills, expertise and experience of managing the pumping of groundwater from the mining shafts were lost. When the government stopped subsidising water pumping activities, the mine managers and their accountants would only pump out the toxic water as long as the mines could operate profitably (ETIMC, 2010). Once the profit margin dropped into the red, pumping water was stopped and shafts were left to flood, thereby rendering them virtually useless for any future operations. A particular case in point were the Aurora Mining Company's operations in the Eastern Basin of the Witwatersrand, where mining shafts started flooding once the government withdrew their funding for pumping operations (ETIMC, 2010).

The governance problem that presented itself was that the existing policy framework for dealing with AMD was not well coordinated. In particular, the delegation of powers between the local, provincial and national government was not properly delineated. In the view of some, the government also came across as reactive instead of proactive (Manders *et al.* 2009). In an effort to tackle the problem more pro-actively, an inter-ministerial committee of experts was appointed to investigate the matter. They reported back to the government in October 2010, but the information was only made public in February 2011. The most significant feature of the report was the vast sums of money required to address the problem. The Trans-Caledon Tunnel Authority (TCTA), a parastatal organisation created at the time of the construction of the Lesotho Highlands Water Project, began to play an important role in attempts to solve the problem.

In 2011, two banking houses funded another report on the anticipated AMD flooding of the central basin of the Witwatersrand and its potential impact on the Johannesburg central business district. The study contested some of the findings of the inter-ministerial committee's report. The winde report warned against over-exaggerating the problem. They also cautioned against spending money with seeming abandon on a project whose estimated costs were escalating alarmingly (Winde *et al.* 2011).

7.3 THE ISSUE OF AMD IN THE COAL MINING SECTOR

The development of South Africa's coal mining industry in the nineteenth century coincided with the height of the British Industrial Revolution when coal-based energy was responsible for driving manufacturing Britain's extensive network of industrial activity, including transport systems with steam-powered ships and railways linking diverse parts of the globe. By the end of the nineteenth century, coal also became an important energy provider for generating electricity in many parts of the world (Freese, 2006). South Africa was a direct beneficiary of these technological developments. The country has limited consistent river water flows that militate against the use of large-scale hydro-electricity projects. This being so, from the outset there was heavy reliance on coal-fired electricity at the diamond fields of Kimberley. In 1879 the demand for energy caused coal mining to begin at Vereeniging on the banks of the Vaal River. And in 1887, shortly after the goldfields of the Witwatersrand opened up, coal mining began at Boksburg (McCarthy, 2011). The steam railway linking the former Transvaal with the Portuguese colonial harbour of Delagoa Bay in Mozambique made it viable in 1894 for the Witbank coalfields to be mined. By the beginning of the twentieth century new power stations were erected where there were substantial coal deposits and sufficient water to begin mining operations.

The downside of coal mining was that as in the case of gold mining, AMD was one of the negative outcomes (McCarthy, 2011). In what has been described as the exploitative phase of mining in South Africa, in the first half of the twentieth century, the objective was essentially to extract the resources as quickly as possible; only later would there be a shift to more sustainable practices. Since the 1970s, South Africa has thus been a major exporter of high-grade coal (Kavalov & Peteves, 2007; Wilson, 2001), with the low-grade yield primarily used locally for electricity production and the manufacture of petroleum from coal.

Initially coal mining was largely conducted by means of shaft-sinking and underground activities. In the 1970s, open-cast mining gained momentum, stepping up production significantly (Wells *et al.* 2009). This led to production and growth in electricity supply, but also implied greater dependence on coal for energy.

Another major problem was that the exact procedures for mine closure were not stipulated. The first guidelines for mine closure were introduced only in 1979. Scant attention was paid to the continued productive use of the land once coal mining had ceased. The South African government lagged significantly behind the prevalent trends in Western countries, such as the United States and Britain, in some cases by as much as twenty years. The consequence was that a serious state of decline was already evident in worked-out coal mining areas of South Africa by the 1970s (Wells *et al.* 2009).

7.3.1 Coal production and the growing demand for electricity

In the transition to a new South Africa in the 1990s, part of the government's objective was to provide more South Africans – especially people from previously disadvantaged communities – with a consistent supply of electricity. The Electricity Supply Commission (Eskom) a parastatal established in 1923 (Bode, 2011), was increasingly used to generate more electricity to meet the demand for power. Although the mining sector experienced a significant increase in the number of new companies staking their claim to conduct mining operations, many of the multinational mining corporations continued with their activities, especially for the purposes of exporting high quality coal.

The downside of the new political dispensation was that coalmining companies, especially in the coalfields of Mpumalanga, started operations that were not in accordance with the legislation controlling the mining industry. This had a negative impact on agriculture in the region because more farmlands were taken up by coalmining activity and water resources, especially those in wetlands, were increasingly subjected to AMD. Moreover, apart from groundwater contamination, a porous water-rich region in the Klipstapel and Lake Chrissie area of Mpumalanga was opened for prospecting. This jeopardised the quality of water flowing into the Upper Vaal River, one of the primary sources feeding the Upper Vaal River system, an important source of natural water supplies for Gauteng and the irrigation of farming areas in the arid and semi-arid interior of South Africa.[3]

The stark consequences of AMD are evident in the Emahlaleni Municipal area, where Witbank has in recent years experienced major problems with decanting acid mine drainage into the catchment of the Olifants River (McCarthy, 2011). The city is home to a number of large industries and is currently a major contributor to the national electricity supply grid. Further downstream, AMD has also affected rural communities as well as platinum mining operations in the Steelpoort Valley. Furthermore, the contaminated water poses a threat to rivers flowing from the Drakensberg Escarpment into the Lowveld and passing through the Kruger National Park, before flowing into neighbouring Mozambique.

Meanwhile, there have been comprehensive measures to open up new mining areas (Jeffrey, 2005). The planning and building of power stations has not gone unnoticed.

7.3.2 Addressing the problem?

The critical areas currently earmarked for combating AMD are Gauteng's Witwatersrand region (especially the Western Basin) and the vicinity of Emahlaleni in Mpumalanga.

[3]There is a map showing the extent to which land has been set aside for the purposes of exploring for coal mining in Mpumalanga Province. According to Dr. Koos Pretorius, an activist in the province, coal mining in many parts of the province could seriously jeopardise the water supplies in the Upper Vaal River catchment. The region is a key provider of water to Gauteng Province. See McCarthy (2011).

In the case of Gauteng the media has created a significant public awareness of AMD in places such as the Cradle of Humankind, near the Sterkfontein Caves, a World Heritage site, where the traces of early humankind date back to more than 3 million years (Durand *et al.* 2010). AMD in the Western Basin, close to the mining cities of Randfontein, Krugersdorp and Westonaria remains a problem of critical proportions, especially in view of the potential threat of radioactivity. The Central Basin region, which extends from the worked-out site of Durban Roodepoort Deep Mine, past Johannesburg close to the East Rand city of Boksburg, is densely populated, and AMD flooding can pose a threat to water courses and ultimately also humans resident in the region. Consistent pressure by activists, especially members of the Foundation for a Sustainable Environment (FSE), an NGO under the leadership of activists Ms Mariette Liefferink, in Gauteng, and Dr. Koos Pretorius in Mpumalanga, has kept the issue of AMD alive in the public sphere.

In 2011, the government considered several alternatives. Its planning commission under the responsible minister, Mr. Trevor Manuel, working on projections of South African society in the future, have made a number of recommendations. In December 2011, at the time of the United Nations' COP 17 climate change conference in Durban in December 2011, the government announced that it was working pro-actively on the issue. In January 2012 the finance minister, Pravin Gordhan, told the legislative assembly during his annual budget vote that an amount of R17 million had been set aside for an on-going feasibility study to be spread over a number of years to address the problem of AMD (Creamer, 2012). The following month, in his annual state of the nation speech, President Jacob Zuma told the country that an amount of R248 million would be available for addressing the issue of AMD in the next two years (Zuma, 2012).

7.4 CONCLUSION

The AMD problem remains formidable. Large sums of money must be invested in Mpumalanga's infrastructure to prevent toxic acidic water from penetrating drinking water supplies in urban and rural settlements (Blaine, 2011). The government has meanwhile tasked the parastatal Trans-Caledon Tunnelling Authority (TCTA) to work on strategies aimed at addressing the AMD threat. There are extensive consultations with stakeholders and the public debates by experts are transparent, but decisive steps still have to be taken (Email circular, 2012).

At the same time, government has to play a balancing game. There is a great deal of political sensitivity about ownership of the mining sector. "Rumours" and public statements by some political leaders have hinted at the government "nationalising" mines; such talk is detrimental to the general stability of the population at large. A rumour to this effect in 2002 triggered a phase of foreign investment withdrawal exceeding R56 billion (Anon, 2009). The issue was exacerbated in 2009 when the then ANC Youth League president, Julius Malema, came out strongly in favour of government nationalising the mines to secure redress for all the country's people (Brown, 2009).

In addition, the management of the country's major mining companies appear to be reluctant to participate in talks with government about seeking solutions to the problem of AMD. In terms of the current legislation, companies have to plan for the rehabilitation of sites after mine closure. Up to the present, there has been little, if any, tacit support for the government's proposals. With consensus not yet reached, there does not seem to be any likelihood of action in the immediate future. Consistent statements by senior parliamentarians (Seccombe & Mokeli, 2012) to the effect that nationalisation of mines is not part of the government's plans (Staff reporter, 2012) is perhaps borne out in proposals on seeking collaborative solutions to AMD.

At present, government appears to be loath to enforce policies of "polluter pays" and accountability in a mining industry where a strong competitive spirit tends to lead to unprincipled and irresponsible behaviour in combating AMD and other pollution-related issues. For example, there are many examples of mining companies that continue their operations despite the fact that they do not possess water licences issued by the Department of Water Affairs. Water licensing is a prerequisite for any mining activity in South Africa (Van Rensburg, 2012; Tempelhoff, 2012a). In 2011 the first legal charges were laid by an NGO, the Federation for a Sustainable Environment (FSO), against a coal mining company in Mpumalanga for unlawful prospecting and mining operations that contravened the country's environmental legislation (Blaine, 2011). As law enforcement officials closed in on more than 400 mining and prospecting companies in April 2012, the Dutch owned company, Anker Coal, and Mineral Holdings in Mpumalanga were the first to be found guilty and fined for contravening existing South African legislation (Seccombe, 2012; Blaine, 2012).

The South African government, as is the case in most democratic societies whose governments have the economic development of the country's people at heart, tends to be too lenient in enforcing crucially important basic environmental principles and policies. The major challenge is to create wealth and eliminate poverty. Consequently, the role of the mining industry is bolstered by the rising demand for mineral resources and the diminishing available supply. At the same time government, as seems to be the case in many other African economies where minerals play an important role in securing growth, is not expected to resort to a "Dutch disease" strategy (Solomon, 2011), of using mining to realise its development objectives over the immediate short term.

The extent to which society can be harmed by insufficient environmental safeguards became evident when the potable water supplies of Carolina, a small mining town of about 20,000 residents, became contaminated with acid mine drainage in January 2012.

The residents were unable to use their tap water, and clean supplies had to be brought in by tankers as the municipality and department of water affairs grappled with plans to clean up the municipal water purification plant. By May 2012 the problem had not yet been properly resolved. Consequently, there were protests by local residents causing significant damage to public buildings and the equipment of contractors who were at the time repairing the water reticulation system in parts of the town (Figure 7.2).

Figure 7.2 After being without proper water supply for more than four months angry residents (mostly youths) in the Mpumalanga town of Carolina on 24–25 May 2012 started protesting and causing considerable damage. *Source*: Photograph JWN Tempelhoff 25 May 2012.

Provisional estimates suggested it would cost at least R600 million to restore the potable water supply of Carolina. Meanwhile the whole population of a town had to experience the discomfort of being unable to properly use the municipal water supply for basic hygiene and sustenance (Blaine, 2012; Tempelhoff, 2012, 2012c).

REFERENCES

Akcil C. and Koldas S. (2006). Acid mine drainage (AMD): causes, treatment and case studies. *Journal of Cleaner Production*, **14**, 1139–1145.

Anon. (2009). More to Malema's remark? *Business Day*, 14 July, at http://www.businessday.co.za/Articles/Content.aspx?id=75623 (Accessed 2012.02.10).

Archer E. (2010). Climate Change and Water. In Council for Scientific and Industrial Research (CSIR), *CSIR perspective on water in South Africa – 2010* (Report No. CSIR/NRE/PW/IR/2011/0012/A), pp. 24–25.

Blaine S. (2011). 'End of an era' as miner faces environmental charges. *Business Day*, 23 February at http://www.businessday.co.za/Articles/Content.aspx?id=135095 (Accessed 2012.05.03).

Blaine S. (2011). Witbank needs R250m to replace old pipes. *Business Day*, 11 October, at http://www.businessday.co.za/Articles/Content.aspx?id=155633 (Accessed on 2012.02.10).

Bode C. (2011). The History of Hydropower in South Africa. (Masters Assignment, North–West University, Potchefstroom).

Brown K. (2009). ANC cool on allies' demand to seize mines. *Business Day*, 6 July, at http://www.businessday.co.za/Articles/Content.aspx?id=75008 (Accessed 2012.02.10).

Buranelli V. (1979). *Gold: An Illustrated History*. Red Dembner Enterprises, USA, (no place of publication indicated).

Coetzee C. B. (ed.) (1976). *Delfstowwe van die Republiek van Suid-Afrika: Handboek, 7* (Mining resources of the Republic of South Africa: Handbook 7). Government Printer, Pretoria.

Cousins C. A. (1976). Platinumgroep metale (Platinum group of metals). In: Delfstowwe van die Republiek van Suid-Afrika: Handboek, 7 (Mining resources of the Republic of South Africa: Handbook 7), C. B. Coetzee (ed.), Government Printer, Pretoria, pp. 93–98.

Creamer T. (2012). Water affairs confirms movement on long-term AMD study. *Engineering News*, 27 January, at http://www.engineeringnews.co.za/article/water-affairs-confirms-movement-on-long-term-amd-study-2012-01-27-1 (Accessed 2012.02.09).

De Jager F. S. J. (1976). Steenkool (Coal). In: Delfstowwe van die Republiek van Suid-Afrika: Handboek, 7 (Mining resources of the Republic of South Africa: Handbook 7), C. B. Coetzee (ed.), Government Printer, Pretoria, pp. 363–404.

Durand J. F., Meeuvis J. and Fourie M. (2010). The threat of mine effluent to the UNESCO status of the Cradle of Humankind World Heritage Site. *TD The Journal for Transdisciplinary Research in Southern Africa*, **6**(1), July, 73–92.

Email circular. (2012). Steve Horak (Digby Wells Environmental), Background information document for the Witwatersrand goldfields AMD project EIA, 9 February.

Expert Team of the Inter-Ministerial Committee (ETIMC). (2010). Mine water management in the Witwatersrand gold fields with special emphasis on acid mine drainage: report to the inter-ministerial committee on mine drainage (December).

Freese B. (2006). Coal: A Human History. (Arrow Books, London).

Haarhoff J. and Tempelhoff J. W. N. (2007). Water supply to the Witwatersrand (Gauteng) 1924–2003. *Journal for Contemporary History*, **32**(2), December, 95–114.

Hammerbeck E. C. I., Fourie G. P., Schoeman A. A. and Schoeman J. J. (1976). Yster (Iron). In: Delfstowwe van die Republiek van Suid-Afrika: Handboek, 7 (Mining resources of the Republic of South Africa: Handbook 7), C. B. Coetzee (ed.), Government Printer, Pretoria, pp. 241–252.

Jeffrey L. S. (2005). Characterization of the coal resources of South Africa. *Journal of the South African Institute of Mineral Metallurgy*, **105**(2), February, 95–102.

Kavalov B. and Peteves S. D. (2007). The Future of Coal. DG JRC Institute for Energy, European Commission, Petten.

Klein N. (2011). Democracy born in chains. *Naomi Klein*, 13 February, at http://www.naomiklein.org/articles/2011/02/democracy-born-chains (Accessed 2012.02.13).

Manders P., Godfrey L. and Hobbs P. (2009). Acid mine drainage in South Africa. In CSIR Briefing Note, 02, August.

Masiza Z. (1993). A chronology of South Africa's nuclear program. *Nonproliferation Review*, **1**(1), January, 34–53.

McCarthy T. S. (2011). The impact of acid mine drainage in South Africa. *In South African Journal of Science*, **107** (5/6), Art. #712, 7 pages. doi: 10.4102/sajs.v107i5/6.712.

Pogue T. E. (2006). The Evolution of Research Collaboration in South African Gold Mining: 1886–1933. PhD, University of Maastricht, The Netherlands.

Research Channel Africa. (2011). South Africa's water sector in 2011 (Craemer Media).

Seccombe A. (2011). Crackdown on errant mining groups: department of mineral resources cracking down on errant mining companies. *Business Day*, 20 April, at http://www.businessday.co.za/Articles/Content.aspx?id=140786 (Accessed 2012.05.03).

Seccombe A. and Mkokeli S. (2012). Mining indaba: mining report dismisses nationalization of mines. *Business Day*, 7 February, at http://www.businessday.co.za/Articles/Content.aspx?id=164328 (Accessed 2012.02.10).

Solomon M. H. (2011). A conceptual approach to evaluating the political economies of mining in Africa and the sector's contribution to economic diversification. *Journal of the South African Institute of Mineral Metallurgy*, **111**(7), July, 475–492.

Stoch E. J. and Winde F. (2010). Threats and opportunities for post-closure development in dolomitic gold mining areas of the West Rand and Far West Rand (South Africa) – a hydraulic view Part 3: planning and uncertainty – lessons from history. *Water SA*, **36**(1), January, 83–88.

Staff reporter. (2012). Nationalisation not ANC policy – Zuma. *Business Day*, 10 February, at http://www.businessday.co.za/articles/Content.aspx?id=164709 (accessed 10 February 2012).

Tempelhoff E. (2012). Departement help Caroline se suur water reinig. *Beeld*, 22 February, at http://www.beeld.com/Suid-Afrika/Nuus/Departement-help-Carolina-se-suur-water-reinig-20120220 (accessed 3 May 2012).

Tempelhoff E. (2012a). M'langa maak stem dik oor myngroep. *Beeld*, 13 February, at http://www.beeld.com/Suid-Afrika/Nuus/Mlanga-maak-stem-dik-oor-myngroep-20120213 (accessed 15 February 2012).

Tempelhoff E. (2012b). Net suur water in hul pype. *Beeld*, 15 February, at http://www.beeld.com/Suid-Afrika/Nuus/Net-suur-water-in-hul-pype-20120215 (accessed 3 May 2012).

Tempelhoff E. (2012c). Net suur water in hul pype. *Beeld*, 15 February, at http://www.beeld.com/Suid-Afrika/Nuus/Net-suur-water-in-hul-pype-20120215 (accessed 3 May 2012).

Tempelhoff J. W. N. (2001). Time and the river: observations on the Vaal River as source of water to the Witwatersrand 1903–24. *Historia*, **46**(1), May, 247–270.

Tempelhoff J. W. N. (2006). Omgewingslagoffers of armblankes? Water en welvaartskepping in die era van Hertzog (1924–1939). Suid–Afrikaanse Akademie vir Wetenskap en Kuns, Pretoria.

Thorius T. (2004). The effect of Grootvlei mine water on the Blesbokspruit. MSc, Rand Afrikaans University [Univklm,. ersity of Johannesburg].

Turton A. (2012). Mining, fracking and acid mine drainage in South Africa – a beginner's guide. *Karoo Space Magazine* at http://www.karoospace.co.za/karoo-space-magazine/talking-point/118-mining-fracking-and-acid-mine-drainage-in-south-africa-a-beginners-guide (accessed 13 February 2012).

Van Rensburg D. (2012). Appèl teen MWS-lisensie verwerp. *Beeld*, 9 January, at http://152.111.1.88/argief/berigte/beeld/2012/01/09/B1/14/dvrFirstUranium.html (accessed 13 February 2012).

Wells J. D., Van Meurs L. H., Rabie M. A., Joubert G. F., Moir F. and Russel J. (2009). Terrestrial minerals. In: Fuggle and Rabie's Environmental Management in South Africa, H. A. Strydom and N. D. King (eds.), Second edition, Juta Law, Cape Town, pp. 513–578.

Wilson F. (2001). Minerals and migrants: how the mining industry has shaped South Africa. *Daedalus*, **130**(1), Winter, 99–121.

Winde F. and Sandham L. A. (2004). Uranium pollution of South African streams – an overview of the situation in gold mining areas of the Witwatersrand. *GeoJournal*, **61**(2), 131–149.

Winde F. and Stoch E. J. (2010a). Threats and opportunities for post-closure development in dolomitic gold mining areas of the West Rand and Far West Rand (South Africa) – a hydraulic view Part 1: Mining legacy and future threats. *Water SA*, **36**(1), January, 69–74.

Winde F. and Stoch E. J. (2010b). Threats and opportunities for post-closure development in dolomitic gold-mining areas of the West Rand and Far West Rand (South Africa) – a hydraulic view Part 2: opportunities. *Water SA*, **36**(1), January, 75–82.

Winde F., Stoch E. J., Erasmus E. and Schrader A. (2011). Desktop Assessment of the Risk for Basement Structures of Buildings of Standard Bank and ABSA in Central Johannesburg to be Affected by Rising Mine Water Levels in the Central Basin (Vol. I of III). North–West University, Potchefstroom, May.

Zuma J. (2012) State of the nation 2012: full address. *Business Day*, 10 February, at http://www.businessday.co.za/articles/Content.aspx?id=164635 (accessed 10 February 2012).

Chapter 8

Safety of lead water pipes: history and present

Heikki S. Vuorinen, Petri S. Juuti and Tapio S. Katko

8.1 INTRODUCTION AND OBJECTIVES

Lead was one of the first metals known to ancient people already thousands of years ago. Its excellent properties for use as water piping made it a very convenient material across the ancient Roman world (Nriagu, 1983a:240–245). A common topic of speculation has been the possible role of widespread lead poisoning, especially among Roman aristocrats, in the decline of the Roman Empire (Gilfillan, 1965; Nriagu, 1983a:407–415, 1983b; Scarborough, 1984; Needleman & Needleman, 1985; Aufderheide, 1993).

Lead is a substance found in inhaled air, food and drinking water, which seldom if ever originates from natural sources. It is a toxic metal, which accumulates in the human body. However, the symptoms of lead poisoning are diverse and none of them alone is a sure indication of lead poisoning. The symptoms include, for instance, lethargy, abdominal pain and headache. Various gastrointestinal problems and symptoms resulting from neuropathy, encephalopathy and nephropathy are also common. Behavioural changes and cognitive problems are also linked to lead poisoning.

The aim of this review is to examine how the toxicity of lead in drinking water was regarded in different periods, from antiquity to modern times. The paper is based on ancient literary sources and special studies on the history of public health and different aspects of health hazards connected with the use of lead. English translations of the ancient texts are from the referred books except for Palladius whose words were translated by the first author.

From decision-making point of view the paper uses path dependence theory. The path dependence theory contends that decisions made in the past are very likely to have long-term impacts on forthcoming events by binding, limiting, or postponing alternative options. As such, path dependence is linked to history and futures research and their interaction. Path dependence has been offered as an alternative analytical perspective for economics that leads to revolutionary reformulation of the neo-classical paradigm. (Kaivo-oja *et al.* 2004; Rajala, 2009:33).

8.2 LEAD IN ANTIQUITY

Lead was widely used in water supply systems in antiquity because it was well known and pipes could be easily made from it. However, lead was considered hazardous to health already in antiquity and was therefore not recommended for water distribution pipes. The first description of a lead poisoning is attributed to Nicander of Colophon in his *Alexipharmaca* from the second century BC (Nicander, 1953: 98–99, 134–135).

The ancient Greeks and Romans were quite aware of some of the risks of poisonous compounds in water. The dangers inherent in water coming from hills and mountains, where mining was carried out, did not escape the notice of a keen ancient observer. The Hippocratic author of *Airs, Waters, Places* wrote in the second half of the fifth century BC: *The next worst will be those whose springs are from rocks – for they must be hard – or from earth where there are hot waters, or iron to be found, or copper, or silver, or gold, or sulphur, or alum, or bitumen, or soda. For all these result from the violence of heat. So from such earth good waters cannot come, but hard, heating waters, difficult to pass and causing constipation.*" (*Airs, Waters, Places*. 7).

Educated people engaged in construction were also aware of the dangers. The author of the first known book on architecture, Vitruvius, expressed these doubts in the late 1st century BC: "*But when gold, silver, iron, copper, lead and the like are mined, abundant springs are found, but mostly impure. They have the impurities of hot springs, sulphur, alum, bitumen; and when the water is taken into the body and, flowing through the vessels, reaches the muscles and joints, it hardens them by expansion. Therefore the muscles swelling with expansion are contracted in length. In this way men suffer from cramp or gout, because they have the pores of the vessels saturated with hard, thick and cold particles.*" (Vitruvius *De Architectura*. 8.3.5).

In addition to Vitruvius, a number of authors touched on different aspects of lead and water during the Roman Era. These included authors of treatises on agriculture like Columella (1st century AD) and Palladius (4th century AD); Plinius (23/24–79 AD), the famous encyclopaedist; and Frontinus (ca. 40–103 AD), the curator of the water works of Rome.

Doubts concerning the use of lead pipes were expressed, and use of ceramic pipes in their stead was recommended (Vitruvius *De Architectura*. 8.6.10–11, Columella *Rei Rusticae* 1.5.2, Plinius *NH*. XXXI.31.57. Palladius *Opus Agriculturae*. 9.11). However, it seems that although ceramic pipes were also used, water was routinely distributed through lead pipes (Figure 8.1), as revealed by both written sources (Vitruvius *De Architectura*. 8.6.1, 4–6; Frontinus. 25.2, 27.3, 29.1, 30.1, 39–63, 105.5, 106.3, 115.3, 118.4, 129.4–6) and archaeological remains (Bruun, 1991:124–127; Hodge, 1992:307–315). Thus, at least in theory, it was possible to get lead poisoning through water use. Yet, there are two reasons to believe that exposure to lead through water was quite minimal, as pointed out by A. Trevor Hodge (Hodge, 1981, 1992:308). Firstly, due to geological reasons, most Central and South European ground waters are rich in calcium (contrary to the situation in Fennoscandia), and consequently calcium carbonate formed a coat on the inner pipe surface separating the lead from water. Secondly, because of the constant flow, the contact time between water and the lead of the pipe was too short for significant lead contamination.

Vitruvius, however, was quite explicit in his recommendation for the use of ceramic water pipes: "Etiamque multo salubrior est ex tubulis aqua quam per fistulas, quod per plumbum videtur esse ideo vitiosum, quod ex eo cerussa nascitur; haec autem dicitur esse nocens corporibus humanis. Ita quod ex eo procreatur, si id est vitiosum, non est dubium, quin ipsum quoque non sit salubre. Exemplar autem ab artificibus plumbariis possumus accipere, quod palloribus occupatos habent corporis colores. Namque cum fundendo plumbum flatur, vapor ex eo insidens corporis artus et inde exurens eripit ex membris eorum sanguinis virtutes. Itaque minime fistulis plumbeis aqua duci videtur, si volumus eam habere salubrem. Saporemque meliorem ex tubulis esse cotidianus potest indicare victus, quod omnes, et structas cum habeant vasorum argenteorum mensas, tamen propter saporis integritatem fictilibus utuntur." (Vitruvius. *De Architectura*. 8.6.10–11.)

Figure 8.1 Lead pipes were commonly used water distribution lines in antiquity, as in the Roman Aquae Sulis, situated in modern Bath, Britain.
Source: Photograph Tea Lallukka.

English translation of the above text: "*Again, water is much more wholesome from earthenware pipes than from lead pipes. For it seems to be injurious by lead, because white lead is produced by it; and this is said to be harmful to the human body. Thus if what is produced by anything is injurious, it is not doubtful but that the thing is not wholesome in itself. We can take example by the workers in lead who have complexions affected by pallor. For when, in casting, the lead receives the current of air, the fumes from it occupy the members of the body, and burning them thereupon, rob the limbs of the virtues of the blood. Therefore it seems that water should not be brought in lead pipes if we desire to have it wholesome. Our daily table may show that the flavour from earthenware pipes is better, because everybody, even when they pile up their tables with silver vessels, for all that, uses earthenware to preserve the flavour of water.*"

Pliny the Elder, for his part, made only a few short comments about the materials to be used in water pipes: "*For the rest, the best way for water to be brought from a spring is in earthenware pipes two fingers thick.*" (Plinius. *NH*. XXXI.31.57) The passage by Columella is also a short one: "*... this rain water is after all most suitable to the body's health, and is regarded as uncommonly good if it is conveyed through earthen pipes into a covered cistern.*" (Columella. *Rei Rusticae* 1.5.2.) Palladio most probably took cue from Vitruvius when he warned against using lead pipes (Nriagu, 1983a:330). "*But, because it is healthful and advantageous, water is conveyed by clay tubes, ... Ultimate proof that lead pipes make water harmful is that ceruse, which is harmful to the human body, is a product of the attrition of the lead.*" (Palladius, 1975. *Opus Agriculturae*. 9.11.)

8.3 TOWARDS MODERN TIMES

Frontinus, the water commissioner, stated that a water system needed constant maintenance to function efficiently (Frontinus, 116–123). From the 4th century onwards the interest for piped water diminished and the maintenance of water systems declined and ceased. Although some water systems survived from antiquity to the Middle Ages, especially around the Eastern Mediterranean region, water supply became more locally focused, and one by one the systems built during antiquity fell into disuse.

The Hippocratic thesis *Airs, Waters, Places*, cited before, has been a celebrated medical textbook, and the ideas expressed in it had a definite influence later in antiquity, the Middle Ages and modern times (Miller, 1962). The famous physician Galen (129–c. 216 A.D) wrote a commentary on *Airs, Waters, Places*, and it was among the few Hippocratic treatises to be translated into Latin already in late antiquity (Miller, 1962; Jouanna, 1999:356, 361, 375, 478).

Throughout centuries, evidence of the toxic effects of lead accumulated, but many water supply distribution systems in the world nevertheless used lead pipes. Concerns about lead exposure from leaden plumbing systems have been expressed since the 16th century. Despite several reports of waterborne plumbism (chronic lead poisoning) in the 19th century, the use of lead in plumbing systems continued and there are a lot of examples of lead pipe use in the 19th and early 20th century from across the world.

During the 19th century chemists placed special emphasis on the quality of water and leaching of lead from pipes. Simply put, they considered soft water dangerous for health besides making the pipes less durable, and hard water safe when using lead pipes to deliver water (Troesken, 2006:123–140). This doctrine was known in Finland in the late 19th century when the construction of the first modern water system started in the 1870s in Helsinki, the capital of the Grand Duchy (Lillja, 1938:295–298). Because the water taken from the Vantaa River was soft, the use of lead pipes was suspected to be dangerous. Experiments were carried out with the water and lead pipes, and it was noticed that too much lead dissolved into the water. Consequently, it was decided that lead pipes should preferably be avoided. Other Finnish cities followed the practice and used lead pipes only for joining cast-iron pipes (Figure 8.2) and as pipe material in some rare cases, for instance, in Helsinki.

Figure 8.2 Making a connection in an old steel pipe using molten lead in Tampere in 1998. *Source*: Photograph P. Juuti.

The health hazards of lead pipes became increasingly well known during the 19th and 20th centuries (Hernberg, 2000; Needleman, 2004; Burnham, 2005; Troesken, 2006:107–112). Much of the early evidence on lead-related health effects dealt with occupational exposure, while later on the environmental hazards from lead were also realised. The major health threats from lead were identified as occupational exposure and children's exposure to old plasters and paints containing lead. International recommendations for the maximum healthy level of lead in drinking water fell during the last half of the 20th century. The 1958 WHO *International Standards for Drinking-water* recommended a maximum allowable concentration of 0.1 mg/litre for lead, the 1963 International Standards lowered this to 0.05 mg/litre, the 1971 International Standards increased it back to 0.1 mg/litre, the 1984 Guidelines lowered it again to 0.05 mg/litre, and the 1991 Guidelines lowered the preferred value to 0.01 mg/litre, which was also the guideline value in 2004 and 2011. (WHO, 2004:392–394, 2011: 383–384).

After the Second World War it was noted that old lead pipes can expose people to elevated lead concentrations in water. However, replacement of the old lead pipes, which are still part of many old water supply systems around the world, has been considered expensive and economically unfeasible at least in the short-term. No wonder then that the World Health Organization (WHO) issued the following statement in 2004: "*Owing to the decreasing use of lead-containing additives in petrol and lead-containing solder in the food processing industry, concentrations in air and food are declining, and intake from drinking-water constitutes a greater proportion of total intake. Lead is rarely present in the tap water as a result of its dissolution from natural sources; rather, its presence is primarily from household plumbing systems containing lead in pipes, solder, fittings or the service connections to homes. The amount of lead dissolved from the plumbing system depends on several factors, including pH, temperature, water hardness and standing time of water, with soft, acidic water being the most plumbosolvent.*" (WHO, 2004:392). This statement was repeated almost verbatim by WHO in 2011 (WHO, 2011:383).

8.4 DISCUSSION

Lead was considered a convenient, almost ideal, material for water pipes already in antiquity, and was used until the 21th century in spite of the accumulating evidence of related health hazards. Path-dependence in the use of lead in water pipes is evident. Once the decision to use lead pipes had been taken, it was hard to reverse the development path. Only from the late 19th century onwards the development of modern states and pressure groups together with the progress of science and the mass media led to a situation where health issues became a serious argument against the use of lead for water pipes. The example of Finland shows how use of lead water pipes was avoided because industrialisation and urbanisation occurred late enough, compared to the majority of industrialised countries, and adequate "scientific" knowledge existed.

In fact, the history of lead in drinking water resembles that of leaded petrol, another health hazard. The latter was introduced in the 1920s after a heated debate in spite of the fact that public health experts and the public were aware of the dangers of leaded petrol (Rosner & Markowitz, 1985). Once the decision was made, leaded petrol was used for over half a century before it was banned and replaced by other, hopefully less dangerous, fuels.

In many modern towns old pipes made of lead are still in place and the lead contents of drinking water are high. In the U.S. of the late 1970s and 1980s there were concerns about the metals in drinking water. Related studies in Seattle, Washington; Boston, Massachusetts; and Portland, Oregon showed that the major culprit was lead (Trussel, 2006). At that time some communities still used lead service pipes, and copper pipes were still joined with a solder containing 50 per cent lead. In 1986, Congress banned the use of lead materials in drinking water systems, which later led to the Lead and Copper Rule (LCR) in 1991 (Trussel, 2006).

According to Kimbrough (2007), exposure to lead can come from: (i) lead service lines, (ii) lead plumbing, (iii) lead solder, (iv) copper pipes, and (v) brass fixtures. Kimbrough noted that while the LCR focused mainly on the first four causes, the last one was not properly treated as a potential source of lead. His study showed that the lead and copper found in plastic pipe systems was due to brass fixtures. Kimbrough suggests that the core concepts of the LCR should therefore be re-examined.

In the 2000s, a lively debate was conducted on lead pipes in the U.S., particularly with regard to public schools and their water fountains. For instance, some schools were ordered to let their drinking fountains run for a while in the mornings before use, and various types of remediation programmes were introduced. In Washington, DC, the Water and Sewer Authority (DC WASA) has offered to replace private lead service lines and galvanised in-house pipes (DC WASA, 2010).

Boyd *et al.* (2008) reported on a field test programme after the implementation of lead remediation programme in Seattle public schools. The water monitoring programme carried out in 2004 showed that the average lead concentration was 21 ug/l compared to the EPA guideline of 20 ug/l. A policy of reducing the levels under 10 ug/l was introduced and a remediation programme was implemented. The findings showed that when pipes and fittings containing very little lead are used, the magnitudes of lead concentrations and their variability will be significantly reduced.

In any case, the current U.S. examples show that lead can still, two thousand years after antiquity, be a health risk and a challenge for tap water in homes and schools. The relevance of the precautionary principle, now commonly accepted in many areas like health and the environment, is also arguable. After lead pipes were widely reintroduced in the 19th century, it has been difficult to apply this principle to tap water. One cannot help but wonder, whether modern societies cannot afford to replace their lead pipes by pipes of better material, or whether it is more a question of the lack of political will?

8.5 CONCLUSIONS

The following conclusions on the use of lead pipes and the related health impacts can be drawn:

(i) Lead was widely used for water pipes in antiquity since it was readily available and easy to convert into pipes.
(ii) Lead was considered a health hazard already in antiquity, and early experts did not recommend it as material for water pipes. Yet, due to the calcium-rich (hard) waters, the presence of lead in distribution networks was not considered a big risk for ancient cities.
(iii) Despite several reports of chronic lead poisoning in the 19th century, the use of lead in plumbing systems continued in many countries and cities.
(iv) In countries like Finland, with naturally soft waters, lead pipes were not taken into use in the late 19th century after experiments showed excessive amounts of dissolved lead in water – a wise strategic decision at the time.
(v) Abandoning of lead pipes is considered unfeasible in the short-term.
(vi) The precautionary principle should be applied.

On the whole, the use of lead pipes is an example of path dependence that the old cities of developed and developing societies still seem to have a hard time eliminating.

Acknowledgements

The comments by Osmo Seppälä, the editor and the peer reviewers are highly appreciated. The financial support from Academy of Finland (Decision no. 135843) is gratefully acknowledged.

REFERENCES

Aufderheide A. C. (1993). Lead poisoning. In: The Cambridge World History of Human Disease, K. F. Kiple (ed.), Cambridge University Press, Cambridge, pp. 820–827.

Boyd Gr., Pierson G. L., Kirmeyer G. J. and English R. J. (2008). Lead variability testing in Seattle public schools. *Journal of AWWA*, **100**(2), 53–64.

Bruun C. (1991). The Water Supply of Ancient Rome. A study of Roman Imperial Administration. *Societas Scientiarum Fennica* (The Finnish Society of Sciences and Letters). Commentationes Humanarum Litterarum 93.

Burnham J. C. (2005). Unraveling the mystery of why there was no childhood lead poisoning. *Journal of the History of Medicine and Allied Sciences*, **60**(4), 445–477.

Columella (1941). Rei Rusticae Lucius Junius Moderatus Columella on agriculture; in Three Volumes; Volume I; with a recension of the text and an English translation by Harrison Boyd Ash. The Loeb Classical Library. Harvard University Press, Cambridge, MA, London, England.

DC WASA (DC Water and Sewer Authority) (2010). Lead Pipes. Visited 7 June, 2010. www.dcwasa.com/lead/

Frontinus Sex. Iulius (1925). De aquaeductu urbis Romae. The Stratagems and the Aqueducts of Rome; with an English translation by Charles E. Bennett; edited and prepared for the press by Mary B. McElwain, The Loeb Classical Library. Harvard University Press, Cambridge, MA, London, England.

Gilfillan S. C. (1965). Lead poisoning and the fall of Rome. *Journal of Occupational Medicine*, **7**, 53–60.

Hernberg S. (2000). Lead poisoning in a historical perspective. *American Journal of Industrial Medicine*, **38**, 244–254.

Hodge A. T. (1981). Vitruvius, lead pipes and lead poisoning. *American Journal of Archaeology*, **85**, 486–491.

Hodge A. T. (1992). Roman Aqueducts & Water Supply. Duckworth, London.

Jones W. H. S. (trans.). Airs, waters, places (1923). In: Hippocrates Volume I, with an English translation by W. H. S. Jones. The Loeb Classical Library. Harvard University Press, Cambridge, MA, London, England, pp. 70–137.

Jouanna J. (1999). Hippocrates. Johns Hopkins University Press, Baltimore.

Kaivo-oja J. Y., Katko T. S. and Seppälä O. T. (2004). Seeking for convergence between history and futures research. *Futures, Journal of Policy, Planning & Futures Studies*, **36**, 527–547.

Kimbrough D. E. (2007). Brass corrosion as a source of lead and copper in traditional and all-plastic distribution systems. *Journal of AWWA*, **99**(8), 70–76.

Lillja J. L. W. (1938). Helsingfors stads vattenledningsverk 1876–1936. (Waterworks of the City of Helsinki 1876–1936). Otava, Helsingfors.

Miller G. (1962). "Airs, Waters, and Places" in history. *Journal of the History of Medicine*, **17**, 129–140.

Needleman H. (2004). Lead poisoning. *Annual Review of Medicine*, **55**, 209–222.

Needleman L. and Needleman D. (1985). Lead poisoning and the decline of the Roman aristocracy. *Echos du Monde Classique Classical Views*, **29**, 63–94.

Nicander (1953). Alexipharmaca. In: Nicander, the Poems and Poetical Fragments, A. S. F. Gow and A. F. Schofield (eds.) (Edited with a translation and notes). The University Press, Cambridge, pp. 94–137.

Nriagu J. O. (1983a). Lead and Lead Poisoning in Antiquity. John Wiley & Sons, New York.

Nriagu J. O. (1983b). Saturnine gout among Roman aristocrats. Did lead poisoning contribute to the fall of the Empire? *The New England Journal of Medicine*, **308**(11), 660–663.

Palladius (1975). Opus Agriculturae. Palladii Rutilii Tauri Aemilian Viri Inlustris. Opus Agriculturae, De Veterinaria Medicina, De Insitione. Bibliotheca Scriptorum Graecorum et Romanorum Teubneriana. B. G. Teubner Verlagsgesellschaft, Leipzig.

Pliny (the Elder) (1963). Natural History; In Ten Volumes; Volume VI; with an English translation by W. H. S. Jones, The Loeb Classical Library. Harvard University Press, Cambridge, MA, London, England.

Rajala R. (2009). Long-Term Development Paths in Water Services – the Case of Finland. Tampere University of Technology, Publication 818, Tampere.

Rosner D. and Markowitz G. (1985). A 'Gift of Go'?: the public health controversy over leaded gasoline during the 1920s. *American Journal of Public Health*, **75**(4), 322–352.

Scarborough J. (1984). The myth of lead poisoning among the Romans: an essay review. *Journal of the History of Medicine*, **39**, 469–475.

Troesken W. (2006). The Great Lead Water Pipe Disaster. The MIT Press, Cambridge, MA, London, England.

Trussel R. R. (2006). Water treatment: the past 30 years. *Journal of AWWA*, **98**(3), 100–108.

Vitruvius (1934). On Architecture; In Two Volumes; Volume II; translated into English by Frank Granger, The Loeb Classical Library. Harvard University Press, Cambridge, MA, London, England.

WHO (2004). Guidelines for Drinking-water Quality. Third Edition, Volume **1**, Recommendations. World Health Organization, Geneva.

WHO (2011). Guidelines for Drinking-water Quality. Fourth Edition, World Health Organization, Geneva.

Section III

From Government to Governance

Section III, From Government to Governance has four chapters.

The first chapter was written by one author, Pietilä P. His key interest is the diversity of water supply and sanitation sectors, especially the roles of municipalities in Europe.

The 2nd chapter was written by two authors, Rusca, M. and Schwartz, K. They take a historical perspective on partnerships in the provision of water supply and sanitation services in developing countries.

The 3rd chapter was written by two authors, Hall D. & Lobina E. They write about the birth, growth and decline of multinational water companies.

The fourth chapter was written by one author, Castro J. E. He focusses on the issues of governance and citizenship in water services, especially in Latin America.

Section III

Chapter 9

Diversity of the water supply and sanitation sector: roles of municipalities in Europe

Pekka E. Pietilä

The tasks and duties of municipalities as well as the division of responsibilities between the central government and municipalities vary greatly within Europe. Typically the responsibility for arranging water supply and sewerage services is by law vested in municipalities. The way water and wastewater services are organised is a highly political decision which reflects the values of the decision makers. In some countries and cities water supply is regarded as such an important basic service that the ownership of the water supply and distribution infrastructure is by law confined to public bodies. Wastewater treatment has wider than local importance, and consequently the central government has financially supported particularly the construction of wastewater treatment plants. Yet, there is no single model or formula for how water and wastewater services should be produced within a single country.

9.1 DIFFERENT TRADITIONS OF MUNICIPAL ADMINISTRATION

The role of local government and its relation to the rest of public administration varies a lot in European countries. Dexia (2002b) classifies 15 older EU countries[1] into three categories: (1) federal states, (2) "regionalised states", and (3) unitary sates. In federal states, Austria, Belgium and Germany, the federated states (Bundesländer in Germany) play an important role in the administration and budget control of local governments. In "regionalised states", Spain and Italy, regions have strong powers similar to federated states including legislative power. In unitary states (Denmark, Finland, France, Greece, Luxembourg, Ireland, the Netherlands, Portugal, Sweden and the United Kingdom), local governments have similar powers, but the degree of decentralisation varies. Most of these 15 countries have two tiers of local government, but some have just one (Austria, Portugal, Finland and Luxembourg), and some even three (France, Ireland, Spain and Italy).

In the twelve most recent EU member states[2], local government structures have undergone changes during the last 20 years since the collapse of the influence of the Soviet Union around 1990 (Dexia,

[1]EU members before the accession of new member states in 2004

[2]Countries which joined EU in 2004 and after

2003). The municipal tier was typically favoured during the early years of transition. People were fed up with centralised administration and wished to implement as wide decentralisation as possible.

In the Nordic countries municipalities have traditionally been and still are strong self-governing units at the expense of regional level administration. The importance of local level administration is clearly illustrated by the share of public sector employees at the local level (Figure 9.1). One of the central pillars of municipalities' self-governance is their own taxation power (Henning, 2001).

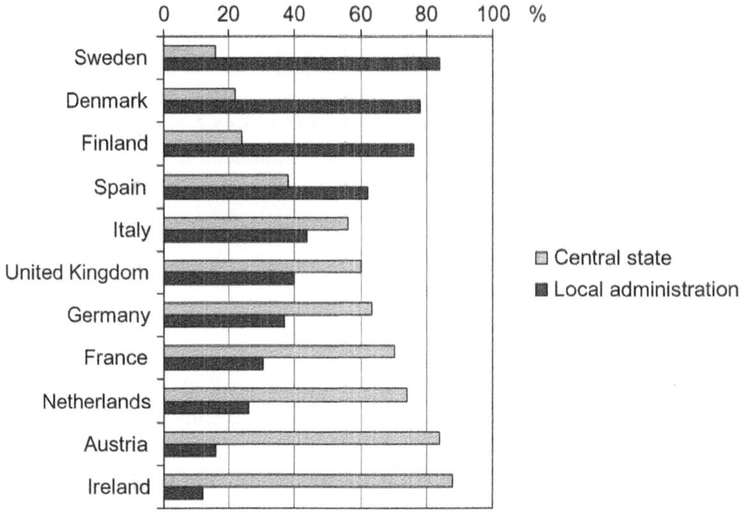

Figure 9.1 Public sector employees in some European countries. *Note*: In this graph employees of the federal states of Germany and Austria are also included in the central government figures.
Source: Dexia 2002a.

In Sweden and Denmark the number of municipalities has reduced dramatically since 1950. In Sweden the number of municipalities has fallen from 1500 in 1950 to 290 in 2010. In Denmark two major municipal reforms were undertaken in 1970 and 2007 reducing the number of municipalities from 1 108 to 98. In Finland the number has not been reduced as much, only from 547 in 1950 gradually to 336 at the beginning of 2011, but the current Cabinet in power since 2011 is planning to cut the number of municipalities dramatically. (Hammarlund, 2002; Kjaer, Hjelmar & Olsen, 2009; Steineke, 2012).

In the Nordic countries, administration at the municipal level, as well as the way water services are organised, has not seen any dramatic changes during the last 150 years. There has been some gradual development, but the basic principles have remained largely the same; municipalities are still rather independent self-administrative units, and they have a central role in water and wastewater service provision. Some apparently big changes have taken place, like the radical reduction in the number of municipalities in Denmark and Sweden. Yet, the responsibilities of municipalities have remained fairly similar. A contrasting example is England, where local level administration has experienced more than one thorough reform during the last 40 years. The water services were also reorganised completely: first ten regional water authorities were formed in 1974 which were then fully privatised in 1989. One explanation for these different development patterns are the different political structures and traditions of UK/England and the Nordic countries. In the UK's basically two-party system radical changes may be expected when the opposition comes to power, while in the multi-party democracies and coalition governments of the Nordic countries political changes are less drastic. In the Netherlands recovery of land from the sea and land drainage have been the backbone of the country's development and still remain vitally important elements, which makes it different from any other European country. Physical necessity rather than local level administration tradition has had the biggest impact on the way water services are managed.

Until the late 1970s, France was a unitary, highly centralised state where the major public tasks were performed by the central state acting through central ministries. The two levels of local government, consisting of 100 départements and 37,000 municipalities were assigned only limited responsibilities. In 1982 the administrative structure was decentralised and responsibilities were transferred from state administration to the départements, and to a lesser degree to the municipalities. As many municipalities have a small population, they have established inter-communal cooperation units for delivering public services such as water supply. In 2000 there were 14,900 single-purpose and close to 2200 multi-purpose inter-municipal entities for the delivery of public services. (Wollmann, 2004).

9.2 FINLAND

The development of centralised water supply, sewerage and wastewater treatment has been remarkably fast in Finland. Still in 1950, only 25% of the Finnish population received piped water into their homes – by 1980 this figure had jumped to 90%. The development took place later than in Sweden, but one should keep in mind the hard times Finland experienced in the middle of the 20th century. (Katko, 1996)

In Finland each municipality typically has its own water and wastewater utility. In addition to municipal utilities, there are various forms of supramunicipal cooperation, in particular for bulk water supply and wastewater treatment. The largest bulk water supplier is Helsinki Metropolitan Area Water Company which supplies water to about one million people (20% of the population of Finland) in eight municipalities. (Pietilä, Katko & Kurki, 2010)

The first wastewater treatment plants for the urban areas of Finland were built in the 1910s, and until the 1950s, only few wastewater treatment plants were in operation. The Water Act of 1961 enabled authorities to set legal requirements and time schedules for water polluters (Katko, Luonsi & Juuti, 2004). This put pressure on municipalities to reduce wastewater pollution loads, and the 1970s, in particular, was an extremely active time of wastewater treatment plant construction.

Finland has seen a gradual development towards larger units in wastewater treatment. In the 1970s the City of Helsinki, with a population of about half a million, had 11 wastewater treatment plants. Since 1994 all the wastewaters of the city have been treated in a single treatment plant which also purifies the wastewaters from five neighbouring municipalities, thus serving a total population of 800,000 people. The longest conveyance distances to the treatment plant are around 50 kilometres which seems to be the maximum viable distance in Finnish circumstances. (HSY, 2011)

Unlike in several other countries, the Finnish state has not subsidised the construction or improvement of larger wastewater treatment plants. The responsibility for financing lies entirely on the municipalities and their water utilities and companies. Some state subsidies have been available for the improvement of water services in rural areas, or to support cooperation between rural municipalities in the water sector.

Finland has long and extensive experience from public-private cooperation in the water supply and sewerage sector, although perhaps not of the type the term is often thought to refer to (i.e. private finance initiative projects). Outsourcing of the services, especially non-core operations, of public water undertakings is very extensive in Finland (Figure 9.2). The contract period is purposely relatively short, such as three years, in order to maintain real competition. It can be argued that in Finland, where water utilities are publicly owned, the elements of private sector competition are far better utilised than for example, in England, where the water sector is 100 percent privatised.

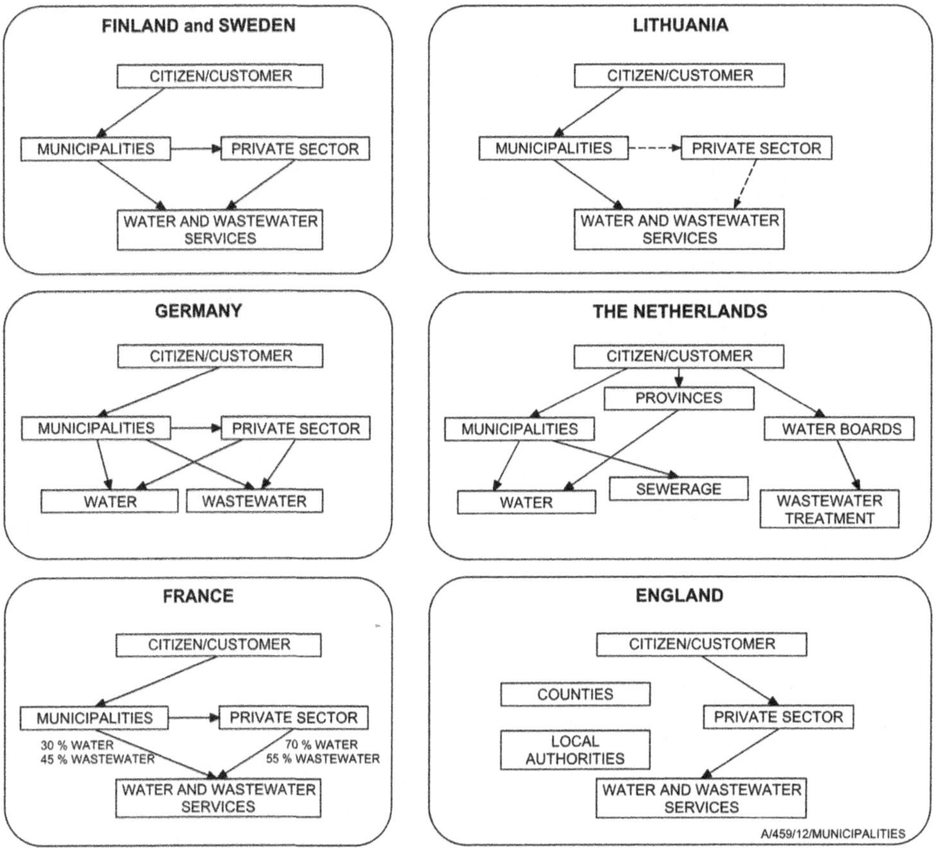

Figure 9.2 Water and wastewater service arrangements in a number of European countries.

In the rural areas of Finland cooperatives have been, and still are, a common means of organising water supply. Water cooperatives have a long history in Finland – the first were established already in the early 1900s (Katko, 1996). Some Finnish water cooperatives are quite large as they serve more than 10,000 people. In many rural municipalities the water cooperative is the only water service organisation which also supplies the centre of the municipality when the municipality has no own water supply unit.

9.3 SWEDEN

During the intensive urbanisation period in 1960–1980, the national government supported financially municipalities that needed to extend their water and sewer networks and build wastewater treatment plants. During the 1970s state support was limited to wastewater treatment plant construction only, but it was available to all municipalities. Since 1980 this kind of state support to water and wastewater facilities has ceased. (Eriksson, 1994)

In Sweden close to 90% of the population live in urban areas, where the coverage of water and wastewater services is 100%. The rest of the population living scattered in rural areas rely on their individual water supply and wastewater disposal systems. Significant expansion of centralised water supply and sewerage systems would be economically unviable due to the long distances and the availability of good water resources and appropriate small scale wastewater treatment solutions. About 1.2 million people, who live in rural areas outside centralised water supply systems, use 400,000 private wells for their water supply. Additionally, there are another 400,000 summer cottages using their own individual water supply. (SGU, 2008).

Since the 1970s per capita consumption of water has decreased clearly: in Stockholm per capita consumption was 500 litres in the early 1970s but had decreased to 319 litres in 2010 (Katko, 1996; Stockholm Vatten, 2010).

The Swedish Public Water and Wastewater Plant Act states that water supply and sanitation as well as the management of storm water are the responsibility of local government (municipalities). According to the Local Government Act (1991:900), municipalities are not allowed to turn a profit from providing basic services such as water supply and sanitation. Earlier, municipalities were allowed to sell their assets and privatise their water and sanitation services, but the new Water Services Act (Lag om allmänna vattentjänster, 2006:412) requires that a municipality (or group of municipalities) must be the majority shareholder in an undertaking that owns water and wastewater infrastructure. Municipalities are, however, allowed to subcontract private companies to operate and maintain their water supply and sanitation services.

One of the very few exceptions, as far as ownership is concerned, was the city of Norrköping, where the city in 1997 established a multi-utility company (Norrköping Miljö och Energi AB) by merging the water and wastewater, energy, solid waste, street and park construction sectors into a municipal company. This company was sold to the energy company Sydkraft, owned by EON Energy AG, in 2001. But when the legislation was changed and no longer allowed private majority ownership of water and wastewater utilities, the municipality bought the shares back in 2005. (Lannerstad, 2003; Pietilä, 2006)

Management contracts are rather new to Sweden. Most of them concern small municipalities, and the contract periods are short, typically three plus three years. Delegated management has been the subject of intensive discussion in Sweden. The concepts proposed, especially by foreign companies, have so far not been considered attractive enough.

9.4 LITHUANIA

During the Soviet rule in Lithuania (1944–1990), water supply and sewerage were the state's responsibilities. The state water and wastewater company had 14 regional subsidiaries for the administration and operation of the systems (Pietilä, 2005). Each subsidiary, again, had subdivisions. In rural areas people used their own wells, as they still do to a large extent. Rural areas also had collective farms which had their own water supply and sewerage systems, and in most cases also wastewater treatment.

After Lithuania regained her independence in 1990, responsibility for public water supply and sewerage was transferred from the state to municipalities (Roman, 2002). Forty-five municipal water companies were established by reorganising the regional state water companies of the Soviet period. Basically each municipality has its own water company, but some companies operate the water supply and sewerage systems of more than one municipality. In addition to these municipal companies, there are also a large number of other smaller water suppliers (cooperatives, residential groups, agricultural companies, municipal companies, other companies, schools, etc.). Stormwater drainage is only seldom the responsibility of water companies but is taken care of by the streets departments of the municipalities.

State water companies were independent units during the Soviet period in the sense that they had all the necessary staff, machinery, equipment and facilities for operating water and wastewater services from the drinking water source to sewer outlets. Municipal water companies inherited this model and were thus, particularly during the early years of municipal companies, self-supporting and managed without the services of private companies. (Kauno vandenys, 2005; Vilniaus vandenys, 2004).

Water consumption in Lithuania is measured with consumer meters. Water use increased steadily until the late 1980s but has dropped dramatically since then, in many areas to one third of what it used to be. This was, on the one hand, advantageous since there was no need to expand the systems as existing capacity was sufficient. But, on the other hand, it is not economical to run over-dimensioned systems. (Pietilä, 2004)

Water tariffs calculated by the municipal company have to be approved by the municipal council (Law on local self-government, 1994). Yet, municipalities are not independent in their tariff setting because tariffs have to be approved also by the National Control Commission for Prices and Energy.

In Lithuania the State budget has been the main domestic source of funding for environmental protection. Since 1999 the funds allocated to municipalities to finance their environmental activities have come from the Privatisation Fund and have been transferred directly to municipalities. So far, the great majority of environmental investments have gone for wastewater systems, in particular improvement of wastewater treatment facilities (Lithuanian Water Partnership, 2002).

In July 2006 a new water supply and wastewater act was approved by the Lithuanian Parliament (Act X-764, Geriamojo vandens tiekimo in nuotekų tvarkymo įstatymas). According to the Act, water supply and wastewater services remain the responsibility of municipalities, and these services have to be provided by public (municipal or supramunicipal) undertakings. Only one water and wastewater undertaking is allowed in each municipality. This requires merging the numerous existing small undertakings with larger units.

9.5 GERMANY

In Germany, as well as most European countries, the pressure to establish public water supply was created by epidemics caused by polluted water. By the 1850s cholera outbreaks had convinced society of the importance of improved water supply and sanitation. In some cases private companies were given concessions to start public water supply or were invited to plan and construct such systems. Yet, water supply was soon taken over by municipalities, and municipalities also established companies for gas and electricity distribution. These municipal companies, Stadtwerke, soon became important revenue sources for municipalities – during the first decade of the 1900s the profits of these companies accounted for 10–16% of municipalities' revenue. (Klenke, 1999).

Private companies showed hardly any interest for centralised sewerage due to the large investments involved. A law enacted in 1900 obliged municipalities to prevent the hygienic hazards caused by wastewater. For several decades, due to health policy reasons, municipalities did not charge fees for treating wastewater covering the costs by municipal taxes (Klenke, 1999).

Already in the second half of the 1800s, the German policy for supplying water to rural areas involved creating large units covering several municipalities. Around 1910, in the district of Württemberg with 457 municipalities, water distribution was implemented by 47 utilities. In North Rhine-Westphalia local groundwater was unavailable because of coal mining, and therefore the municipalities created a joint organisation for water supply (In 1973 this company was renamed Gelsenwasser AG, and it has since grown into a multi-utility company, expanding its activities to the energy and gas sectors; it also operates outside Germany). In the 1960s the coverage and level of water and wastewater services in rural areas was improved with massive support from the state (up to 90% of the costs), and by 1975 already 93% of the population in Germany were connected to public water supply. By 2007 the share of population connected to public water supply has risen to 99%, and that connected to a sewage network to 96%. Even though there are several large regional water and wastewater service providers, the total number of public water or wastewater undertakings in Germany is large. In total, there were in 2008 more than 6200 water supply utilities and enterprises and 6900 wastewater disposal utilities. (Klenke, 1999; Knothe, Kramer & Mohajeri, 2003; Lanz, 2004; wvgw, 2011).

Municipal regulations and the water laws of different German Länder stipulate that drinking water supply is usually, and wastewater disposal always, an obligation of the municipalities as part of their duty to ensure basic living conditions (wvgw, 2011). In Germany water supply has traditionally been organised on a commercial basis, and it has been subject to taxation; the VAT is currently 7% regardless of ownership. The taxation of wastewater treatment depends on ownership. If a wastewater company is privately owned, it has to pay a VAT of 19%, but if publicly owned, the tax rate is zero (Wackenbauer, 2009). In Germany, water and wastewater services have been typically provided by separate organisations as is shown in Figure 9.2. Instead of having water and wastewater services under the same municipal unit, it has been common to have municipal water and gas utilities under a single municipal enterprise (Gas- und Wasserbetriebe). However, a noticeable tendency to combine water and wastewater activities into a single unit has developed in recent years.

In West Berlin, the water and wastewater utilities were merged in 1988 into a single utility, Berliner Wasserbetriebe. After the unification of Germany, the water and wastewater utility of East Berlin was merged with Berliner Wasserbetriebe. In 1999 the company was privatised as the city sold 49.9% of the shares to a consortium of private companies. The reason for the privatisation was not to enable supplying better or cheaper water to Berliners but the severe financial difficulties of the city. (Lanz, 2005; Lanz & Eitner, 2005).

In Hamburg water and wastewater services have been provided by two separate municipal utilities: potable water supply has been the responsibility of Hamburger Wasserwerke (HWW) and sewerage and wastewater service has been provided by Hamburger Stadtentwässerung (HSE). From the beginning of 2006 these two utilities have been under joint management and now operate under the name Hamburg Wasser. However, HWW and HSE remain legally autonomous units. (GWI, 2005; HWW, 2012).

In the 1990s liberalisation and privatisation of water supply was debated in Germany. Some privatisations were carried out and public-private partnerships increased, but, on the whole, the sector is still largely managed by public undertakings. Yet, there have been some clear structural changes over the last 20 years. Water utilities that were organised as municipal departments have been transformed into more independent organisations. In the wastewater sector, the share of services provided by inter-municipal organisations has increased from 4% in 1997 to 28% in 2008. (Wackenbauer, 2009; wvgw, 2011).

In Germany, the water sector functions without formal, external regulation of water tariffs or returns on investment. The leading principle is to cover costs. The great majority of water and wastewater services are provided by municipal undertakings, and their tariffs must be approved by a local government under supervision of a federal state. Even in cases, where the ownership of a company is partly private, the municipality retains substantial influence on the strategic decisions by holding at least a 50.1% majority in the company. These companies operate under private law and are subject to control of anticompetitive practices. Performance comparisons between companies are widely practiced on a voluntary basis. (Wackenbauer, 2009).

9.6 THE NETHERLANDS

In the early Middle Ages people in the boggy peatland of the western part of Netherlands realised that construction of dikes and water drainage were matters that were overwhelming for a single village. They had to join forces and work together for the common interest of safe water management, which resulted in the first water boards. The Water Board Hoogheemraadschap van Rijnland, established as early as

1232, is the oldest water authority still functioning. These Boards are the oldest democratic form of government in the Netherlands. Their number was 3500 in the mid-1800s, and still 2600 in the 1950s. Since then, mergers reduced their number to 26 in 2011. In addition to their original drainage task, Water Boards are now also responsible for wastewater treatment, see Figure 9.2. (de la Motte, 2004; Kuks, 2006; Rijkswaterstaat, 2011).

When public water supply started in the Netherlands in the middle of the 19th century, most companies were under direct private management. Between 1850 and 1920 water services were dominated by private companies but were then gradually taken over by municipalities and operated as public utilities. When the pressure to expand public water supply also to rural areas increased, private companies were in many cases not interested and were bought by the municipalities. Until the 1970s water utilities were predominantly under direct public management but have since become publicly owned companies operating under private law. (Schwartz, 2011).

In 2004 the Parliament of the Netherlands passed a law which prevents any privately owned company from providing drinking water services to the public. However, it does not cover sewerage or wastewater treatment. According to the law, water supply companies or cooperatives have to be 100% publicly owned (Hall, Lobina & de la Motte, 2004). While the water companies themselves remain publicly owned, they contract many services – such as customer service and repairs – out to the private sector.

The number of water companies in the Netherlands has decreased dramatically during the last decades. In the 1950s there were more than 200 water companies, by 1990 the number had been reduced to less than 60, and the mergers continued so that in 2000 just 20 were left, and in 2010 only 10. During the process of municipal utility mergers and creation of regional utilities, many municipalities sold their shares. (Vewin, 2010; Schwartz, 2011).

Water companies in the Netherlands are not subsidised by any public authority which means that users pay the "real price" for their water. Water charges include a fixed and a variable component, and almost all users have a water meter. Municipalities control water prices, there is no national regulator, but the quality of drinking water is overseen by the Ministry of the Environment. (BIPE, 2001; Kuks, 2002).

Sewerage is the responsibility of municipalities in the Netherlands (Kuks, 2006). Because sewers are laid under streets and roads maintained by the municipality, it seems logical to give municipalities a leading role in sewerage. Most systems are combined, but where there are separate stormwater sewers, municipalities are responsible for them as well (Perdok & Wessel, 1998). Private companies are widely used in the construction and maintenance of sewers, so that 80% of the money flows to the private sector (ten Elshof, 2001). The costs of sewerage are covered either by sewerage charges or by property taxes (BIPE, 2001; Kuks, 2002).

Wastewater treatment was slowly introduced in the Netherlands after sewerage systems became more widespread in the 1930s. Since municipalities had constructed the sewer systems, they were also the early initiators of wastewater treatment together with the provinces. The Surface Water Act of 1969 transferred the responsibility for wastewater treatment to Water Boards, but large companies can treat their own effluent or have it done by third parties. The pollution charge is calculated on the polluter-pays principle with industries and institutions paying according to the quantity and quality of their wastewater discharge while households pay a sum based on their metered consumption. (BIPE, 2001; de la Motte, 2005a; Kuks, 2002).

9.7 FRANCE

Since the end of the 18th century, French municipalities gradually took over the responsibility for providing water services, and since the end of the19th century for wastewater collection, but they quickly faced a lack of technical and financial resources for service operation largely due to the small size of the local authorities.

Involving the private sector, based on public works concession contracts, was seen as a way to overcome the problem of lacking resources. (Barraque & Le Bris, 2007).

The operational organisation of water and wastewater services in France is the municipalities' responsibility. A municipality can choose to run the services by itself, or establish a *"syndicat"* with neighbouring municipalities and transfer responsibility to it. The services may be run by a public operator *(régie)* set up by a municipality or a *"syndicat"*, or running can be delegated to a private company. There are three economic types of delegated management: (1) *regie interessee,* where the delegate operates and maintains the assets built by the public authority and receives a proportional fee based on the volume sold, (2) *affermage,* the most common type of delegation, where the delegate operates and maintains the assets built by the public authority, but receives its revenue from the users and pays a fee *(surtaxe)* to the public authority in accordance with the depreciation of the assets, and (3) *concession,* where the delegate builds, operates and maintains the assets and receives its revenue from the users. Delegated management is chosen by less than a third of the organising authorities, but provides about 70 percent of the population with water and treats the wastewater of about 55 percent of the population. (Barraque & Le Bris, 2007).

Private operation of water services has been recently criticised more commonly – a striking example is the re-municipalisation in Paris. Water distribution in Paris was operated by two private companies based on concession contracts, but when the concessions ended, the City of Paris took over the operations in 2009. The example of Paris may have an impact on other local authorities. Many contracts with the private sector are due for renewal or renegotiation in the next few years. An increasing number of local authorities are interested in returning to public management, but not all of them have taken the first step as yet. So far, they have in most cases only threatened to go public in order to renegotiate existing contracts, and lower water rates. (Le Strat, 2010).

9.8 ENGLAND

In England water services were developed until the 20th century predominantly by private companies. In London, the period of private undertakings began in 1581 when the City Corporation granted the use of the first arch of London Bridge to Peter Morris for a term of five hundred years (London's water supply, 1953). During the next centuries, several water companies were established in London, and they competed fiercely for customers so that pipes of several companies could run along a single street.

In a large city like London, such a fragmented water supply structure proved unable to secure reliable and safe water supply to all residents. Since the 1850s attempts were made to bring London's water supply under a single public body. A public utility called Metropolitan Water Board was then established in 1903 to take care of London's water supply. At that time water was supplied to close to seven million people in London. Private water companies were taken over by municipalities also elsewhere in England, so that by 1901 the share of municipal water supply undertakings had reached 90% in larger provincial towns (Hassan, 1998).

The supply of water was followed by the installation of water closets and cesspits. Sewage became an increasing problem for the growing urban population. Rivers were an important source of drinking water, but due to increasing sewage discharges, the quality of the water in the rivers deteriorated. The sewerage systems were, on the whole, constructed and operated by local authorities as part of the public services which they provided at the time. Local authorities obtained their revenue through a levy on every household and commercial, agricultural and industrial property based on the rateable value of the property. (Rees & Zabel, 1998).

From 1900 until 1973 municipalities were predominantly in charge of water and wastewater systems in England, with the exception of about one fifth of the population that was supplied water by non-municipalised private companies (Semple, 1994; de la Motte & Lobina, 2005). In 1915 there were over 2000 water supply undertakings in England and Wales, but this number fell to 1186 by 1945 through regrouping, amalgamations and take-overs. Further mergers occurred in the 1950s as a result of the government policy of concentrating water supply activities. The number of sewerage undertakings, however, remained at around 1400 until 1974 (Rees & Zabel, 1998).

In England and Wales, water services were transferred in 1974 from municipalities to semi-nationalised Regional Water Authorities (RWAs). These new RWAs were based on river catchment areas, and they took over the work previously carried out by 157 water undertakings, 1398 sanitary authorities, and 29 river authorities (Hassan, 1996b). Twenty-eight private companies, which supplied water to 25% of the population, were excluded from the restructuring and continued operations (de la Motte, 2005b).

Then, in 1989, these RWAs were privatised and replaced by 10 private water and sewerage companies which became owners of all water systems and property of the RWAs. However, 29 water companies remained separate from these new 10 large companies, so that 25% of the population in England and Wales continued to have their water (but not wastewater) needs provided by these small companies. The number of these water-only companies decreased due to amalgamations to 12 by 2010. In England, water companies still charge the majority of households on the basis of property tax valuations, with metered charges for commercial users. Metered charges are also applied to new houses, where meters are routinely installed, and a minority of households who have chosen to be metered. (Semple, 1994; Rees & Zabel, 1998; de la Motte, 2005b; Ofwat, 2010).

The prime motivations behind the decision to privatise were political and ideological. During the 1980s the conservative British government had set itself the political objective of privatising the public services wherever practicable, based on its belief that the provision of services is best, and most economical, when carried out by the private rather than the public sector (Semple, 1994). The conservative party was in power in the UK from 1979 until 1997, and during this period virtually all formerly nationalised industries were privatised – telecoms, electricity, gas, water, railways, coal, iron, steel, (Price & Young, 2003). Political authority, and the values and relationships upon which political authority rests, have been replaced by a complex system of regulation, inspection, performance management and audit which has bred the new quasi-profession of auditors, regulators and inspectors that administer and manage (Cooper & Lousada, 2005). Figure 9.2 shows how public bodies are sidelined in water and wastewater services in England.

9.9 DISCUSSION AND CONCLUSION

Even within Europe the ways of organising water and sanitation services vary a lot from one country to another. Different practices are often rooted in different public sector administrative traditions, but may also be influenced by geography and the physical environment. Over the last one hundred years, only a few countries in Western Europe have reorganised their water and sanitation services radically.

Even though private sector participation has been discussed a lot in recent years, water supply and sanitation services are still being provided by the public sector for an overwhelming majority of users globally as well as in Europe. There is a fundamental difference between the basic philosophies of public and private enterprises. Public water undertakings are established to serve the needs of the population, typically in a certain geographical area, and they do not aim to increase substantially their share of the market as private companies do. Of course, private companies also aim to serve the needs of their clients and produce attractive services because winning customers is undoubtedly in their interest. Yet, the prime motive of private companies is to make a profit for their owners.

How water and wastewater services are organised is a highly political decision which reflects the values of the decision makers. The extremes are full privatisation, as in England and Wales in 1989, and full public control without any element of commercialisation as was the case in the former Soviet Union. Yet, it should be noted that both the English privatised and the Soviet public model were highly centralised – even in England, the tariffs had to be approved de facto by the Parliament.

It seems to be typical in the water sector to maintain a strong base of public sector service organisations in the form of enterprises usually owned by municipalities. There has been a general tendency to give more decision making power to these municipal enterprises in order to make them more productive and efficient. Even though decision making is kept in public hands, the services of private companies are widely used.

REFERENCES

Barraque B. and Le Bris C. (2007). Water sector regulation in France. *Journal of Institutional Comparisons*, **5**(2), 4–12.

BIPE (2001). Pricing Water, Comparative Features in Water Service Organisations in France and Europe. BIPE & Stratorg International, Boulogne Billancourt Cedex, France, p. 150.

Cooper A. and Lousada J. (2005). Borderline Welfare. Feeling and Fear of Feeling in Modern Welfare. H. Karnac Books Ltd, London, p. 237.

Dexia (2002a). Local Government Employees in the Fifteen Countries of the European Union. Dexia Editions, Paris, France, p. 39.

Dexia (2002b). Local Financing in the Ten Countries Joining the European Union in 2004. Second edition, Dexia Editions, Paris, France, p. 311.

Dexia (2003). Local Finance in the Twenty Five Countries of the European Union. Dexia Editions, Paris, France, p. 189.

Eriksson Ö. (1994). Trends of organizational development of water supply in Sweden. In: Sound Institutional Strategies for Water and Sanitation Services. Seminar proceedings, J. Hukka, P. Juhola, T. Katko, H. Morange and P. Pietilä (eds.), Tampere University of Technology, Institute of Water and Environmental Engineering, Publication B 59. Tampere, Finland, pp. 18–23.

GWI (2005). Hamburg creates new water player. *Global Water Intelligence*, **6**(9), 12–13.

Hall D., Lobina E. and de la Motte R. (2004). Making Water Privatisation Illegal: New Laws in Netherlands and Uruguay. Public Services International Research Unit, University of Greenwich, London, p. 8.

Hammarlund K. G. (2002). Regional Reforms and Citizen Participation in Sweden: Interim findings of the EUROPUB project. EUROPUB workshop: European Governance and Democrac, Brussels, 14–15 November 2002.

Hassan J. (1998). A History of Water in Modern England and Wales. Manchester University Press, Manchester, UK, p. 214.

HSY (2011). Water services. Viikinmäki wastewater treatment plant. http://www.hsy.fi/en/waterservices/services/wetreatwastewaterefficiently/Pages/default.aspx (accessed 15 February 2011).

HWW (2012). Die Konzernstruktur von Hamburg Wasser. http://www.hamburgwasser.de/konzern.html (accessed 4 December 2011).

Katko T. (1996). *Vettä! – Suomen vesihuollon kehitys kaupungeissa ja maaseudulla* (Water! Evolution of Water Supply and Sanitation in Finland from the mid-1800s to 2000). Vesi- ja viemärilaitosyhdistys, Helsinki, p. 416.

Katko T., Luonsi A. and Juuti P. (2004). Water pollution control and strategies in Finnish forest industries in the 20th century. In: Long-term Development of Water and Environmental Services in Finland, P. Juuti and T. Katko (eds.), From a Few to All. KehräMedia Inc., Finland, pp. 129–149.

Kauno vandenys (2005). Annual Report 2005. Kauno vandenys, Kaunas, Lithuania.

Kjaer U., Hjelmar U. and Olsen A. (2009). Municipal Amalgamations and the Democratic Functioning of Local Councils: The Danish 2007 Structural Reform as Case. European Group of Public Administration (EGPA) conference "The Public Service: Service Delivery in the Information Age", Workshop IV on Local Governance and democracy, Saint Julian's, Malta, 2–5 September 2009.

Klenke M. (1999). *Trink- und Abwassergebuhren in den neuen Bundesländern im Zwielicht politischer Entscheidungen* (Water and wastewater charges in the new German states in the haze of political decision making). Ingenieurburo Klanke, Bruck, Germany. p. 125. ISBN 3-00-005593-2.

Knothe B., Kramer A. and Mohajeri S. (2003). Germany. In: Aqualibrium. European Water Management between Regulation and Competition. European Commission Research Project, S. Mohajeri, B. Knothe, D.-N. Lamothe and J.-A. Faby (eds.), Directorate-General for Research, Global change and ecosystems, Brussels, Belgium, pp. 133–152.

Kuks S. (2002). The Evolution of the National Water Regime in the Netherlands. EUWARENESS Research Project, European Commission 5th Framework Programme. University of Twente, The Netherlands. http://www.euwareness.nl (accessed 15 February 2011).

Kuks S. (2006). The privatization debate on water services in the Netherlands: public performance of the water sector and the implications of market forces. *Water Policy*, **8**(2), 147–169.

Lannerstad M. (2003). Sweden. In: Aqualibrium. European Water Management between Regulation and Competition. European Commission Research Project, S. Mohajeri, B. Knothe, D.-N. Lamothe and J.-A. Faby (eds.), Directorate-General for Research, Global change and ecosystems, Brussels, Belgium, pp. 261–287.

Lanz K. (2004). National Context Report Germany. WATERTIME Research Project. http://watertime.net/docs/WP1/NCR/D10d_Germany.doc (accessed 15 February 2011).

Lanz K. (2005). Germany. In: Water, Time and European Cities – History Matters for the Futures, P. Juuti and T. Katko (eds.), WATERTIME Research Project, http://www.watertime.net, pp. 79–91.

Lanz K. and Eitner K. (2005). Case Study Berlin. WATERTIME Research Project, p. 30. http://www.watertime.net/docs/WP2/D12_Berlin.doc (accessed 15 February 2011).

Le Strat (2010). Paris: an example how local authorities can regain control of water management. Additional Chapter in Publication: Reclaiming Public Water, Transnational Institute & Corporate Europe Observatory. B. Balanyá, B. Brennan, O. Hoedeman, S. Kishimoto, P. Terhorst (eds.), Drukkerij Mittelmeijer, Amsterdam, http://www.tni.org/tnibook/reclaiming-public-water-2 (accessed 4 December 2011).

Lithuanian Water Partnership. (2002). Towards Effective Management of Water Resources in Lithuania. National Report. Vilnius, June-September 2002, p. 44.

London's Water Supply (1953). A Review of the Work of Metropolitan Water Board. Staples Press Ltd, London, p. 368.

de la Motte R. (2004). The Netherlands National Context Report. WATERTIME Research Project. http://watertime.net/docs/WP1/NCR/D10h_Netherlands.doc (accessed 15 February 2011).

de la Motte R. (2005a). Netherlands. In: Water Time and European Cities – History matters for the Futures, P. Juuti and T. Katko (eds.), WATERTIME Research Project, http://www.watertime.net, pp. 134–145.

de la Motte R. (2005b). The UK National Context Report. WATERTIME Research Project. www.watertime.net/docs/WP1/NCR/D10m_UK.doc. p. 31 (accessed 15 February 2011).

Ofwat (2010). Service and delivery – performance of the water companies in England and Wales 2009-10. http://www.ofwat.gov.uk/regulating/reporting/rpt_los_2009-10.pdf (accessed 11 November 2011).

Perdok P. J. and Wessel J. (1998). Country report: Netherlands. In: Institutions for Water Resources Management in Europe, F. N. Correia (ed.), Volume 1. A.A. Balkema Publishers, Rotterdam, The Netherlands, pp. 327–447.

Pietilä P. (2004). National Context Report Lithuania. WATERTIME Research Project. http://watertime.net/docs/WP1/NCR/D10g_Lithuania.doc (accessed 11 September 2011).

Pietilä P. (2005). Lithuania. In: Water, Time and European Cities – History matters for the Futures, P. Juuti and T. Katko (eds.), WATERTIME Research Project, http://www.watertime.net, pp. 123–133.

Pietilä P. (2006). Role of Municipalities in Water Services. Tampere University of Technology, Tampere.

Pietilä P., Katko T. and Kurki V. (2010). *Vesi kuntayhteistyön voiteluaineena* (Water as a lubricant for municipalities' cooperation). Kunnallisalan kehittämissäätiö, tutkimusjulkaisu nro 62. p. 109.

Price C. and Young A. (2003). UK utility reform: distributional implications and government response. In: Utility Privatization and Regulation. A Fair Deal for Consumers? C. Ugaz and C. Price (eds.), Edgar Elgar Publishing, Cheltenham, UK, p. 301.

Rees Y. and Zabel T. (1998). United Kingdom. In: Institutions for Water Resources Management in Europe. Volume 1, F. Correia (ed.), A.A. Balkema, Rotterdam, The Netherlands, pp. 537–670.

Rijkswaterstaat (2011). Water Management in the Netherlands. Ministry of Infrastructure and the Environment, Rijkswaterstaat.

Roman M. (2002). Small community water services in Central and Eastern European Countries – situation, challenges, needs and developments. In: Small Community Water Supplies, J. Smet and C. van Wijk (eds.), IRC International Water and Sanitation Centre, Technical Paper Series 40, Delft, The Netherlands, pp. 48–59.

Schwartz K. (2011). From government-driven to commercially-driven mergers: changing drivers for upscaling in the Dutch water supply sector. *Water Policy*, **13**, 317–325.

Semple A. (1994). Privatised water services in England and Wales. In: Sound Institutional Strategies for Water and Sanitation Services. Seminar proceedings, J. Hukka, P. Juhola, T. Katko, H. Morange and P. Pietilä (eds.), Tampere University of Technology, Institute of Water and Environmental Engineering, Publication B 59, Tampere, Finland, pp. 35–38.

SGU (2008). *Att borra brunn för energy och vatten – en vägledning* (Drilling boreholes for energy and water - guidance). Normförfarande vid utförande av vatten- och energibrunnar. Normbrunn –07. Sveriges geologiska undersökning (SGU), Uppsala, Sweden, p. 36.

Steineke J. (2012). Trends in Nordic Municipal Amalgamations. Nordic Centre for Spatial Development, Stockholm, Sweden.

Stockholm Vatten (2010). *Nyckeltal 2001–2010* (Key figures 2001–2010). http://www.stockholmvatten.se/commondata/rapporter/stockholmvatten/Nyckeltal%20Rapport%202001_2010.pdf (accessed 11 November 2011).

ten Elshof, Erik Association of Netherlands Municipalities. Personal communication 22.9.2001.

Vilniaus vandenys (2004). Annual Report 2004. Vilniaus vandenys UAB, Vilnius, Lithuania, p. 43.

Wevin (2010). Dutch water companies. Drinking water fact sheet 2010. http://www.vewin.nl/english/Dutch%20water%20companies/Pages/default.aspx (accessed 11 November 2011).

Wackerbauer J. (2009). Public or Private Water Management: The German Experience. Second Annual Conference on Competition and Regulation in Network Industries. Brussels, 20 November 2009.

Wollmann H. (2004). Local Government Reforms in Great Britain, Sweden, Germany and France: between multi-function and single-purpose organisations. *Local Government Studies*, **30**(4), 639–665.

wvgw (2011). Profile of German Water Sector 2011. wvgw Wirtschafts- und Verlagsgesellschaft Gas und Wasser mbH. Bonn, p. 103.

Chapter 10

A Historical perspective on partnerships in the provision of water supply and sanitation services in developing countries

Maria Rusca and Klaas Schwartz

10.1 INTRODUCTION

Since the middle of the 20th century water services have been dominated by the public sector. Although other service modalities (private utilities, community-based service provision and co-production of services) have existed for long, service provision through a single public utility has remained the most common service modality. In addition to this dominant service modality, however, the provision of (public) services through "partnerships" has increasingly gained importance. The term partnership has been loosely used to describe a variety of arrangements, encompassing various forms of "co-operation" in the delivery of water services. The argued rationale behind the partnership model lies in the potential to combine comparative advantages of different actors. In their idealized representation partnerships incorporate "a shared sense of purpose, mutual respect and the willingness to negotiate" (Pugh *et al.* 1987 in Lister, 2000:228), "mutual influence, with a careful balance between synergy and respective autonomy [...], mutual respect, equal participation in decision making, mutual accountability and transparency" (Brinkerhoff & Brinkerhoff, 2002:21), and the recognition that all actors can benefit by pooling their resources (UNDP, 2008). These features, however, do not necessarily reflect reality, but rather represent an idealized perspective of what has been defined as a "genuine" partnership (Hauck & Land, 2000:2).

Since the 1970s, different concepts of partnerships have been promoted in the water services sector. Underlying these different concepts are usually ideological considerations and convictions. These convictions mainly pertain to the role of different organizations in the service provision process, the institutions under which services should be provided and the infrastructural approach to service provision. The nature of the partnerships promoted on the basis of these convictions differs in terms of the objectives to be achieved by the partnership, the nature of the partners involved in such a partnership, the guiding principles under which such a partnership takes place, and the level of service that is provided. In this chapter we distinguish four different types of partnerships as they have been promoted over the past decades. We contend that the respective types of partnership reflect different approaches to the water services sector. The distinguished partnerships are public-public twinning projects, public-private partnerships (PPP), water operator partnerships (WOP) and partnerships involving small-scale providers. In this article we link these partnerships to three different time periods.

By distinguishing and contrasting these four types and placing them on a temporal scale, we oversimplify the complexities of the water service sector. Different partnership types have existed simultaneously for many decades. We argue, however, that different partnership types, and their relative prominence, in terms of the degree to which they are being promoted by international financing and development agencies, reflect different perceptions of the water services sector over time (Table 10.1).

Table 10.1 Time periods and types of partnerships.

Time period	Types of partnerships
1945–late 1980s/early 1990s	Twinning Projects
Early 1990s–2003	Public-Private Partnerships
2003–present	Water Operator Partnerships (since 2006); Partnerships of Small-Scale Providers

10.2 PUBLIC-PUBLIC TWINNING PROJECTS

Twinning projects were introduced after the Second World War as a way of developing "intercultural ties" (Boag & McDonald, 2010:3). In the era following the Second World War up to the 1980s, the water services sector in developing countries was strongly characterized by the colonial past and the ensuing postcolonial era. During colonialism, water services systems were designed "to cater primarily to the needs and preferences of the economically strong minority groups" (Nilsson, 2006:381). The conventional piped systems, modeled on practices and preferences of colonial elites, were often too expensive for the indigenous populations. As a result, missions took care of the basic services for indigenous people, while more sophisticated infrastructure services were mostly reserved for the colonial population (Batley, 2006). The presence of two (unequal) systems resulted in a geographical separation of native and colonial areas: "spatial domination was the everyday metaphor for colonial domination" (Chaplin, 1999: 150). In the postcolonial era, water services became a political tool for the newly born states to affirm their recently acquired role and authority. In some cases, such as Nigeria and Malawi, the state extricated private sector and faith based organizations from their role of supplier or incorporated them through public funding (Batley, 2006). The emphasis was on the state ensuring that all people would receive the services previously limited to the colonial elite. Often this was accompanied by a policy of state subsidized tariffs.

Although there is no generally accepted definition of organizational "twinning", it refers to two utilities voluntarily cooperating in order to encourage the "adoption of new procedures, techniques and management systems" (Lariola *et al.* 2000:6). The twinning projects implemented in the 1960s and 1970s should be seen in the above-mentioned postcolonial context. Twinning projects generally adhered to specific principles. First of all, they adhered to a highly standardized and globally applicable service provision ideal. This "modern infrastructural ideal" propagates a relatively high standard of service provision through in-house connections, 24-hour service provision per day and services provided through a single utility. Secondly, the challenges of providing water services were seen as a highly infrastructural and technological problem. By providing a public utility sufficient infrastructure and technological expertise to operate and maintain that infrastructure, it would be able to provide the desired universal and standardized service. Twinning thus had a very strong technological focus. Thirdly, water services were predominantly viewed as a public sector responsibility. Twinning projects thus meant that public utilities "twinned" with public utilities.

Rising government expenditure in combination with poorer than anticipated economic performance inspired questioning of the actual effectiveness of large public bureaucracies in many countries (Hood, 1991; Barzelay, 2001; Minogue, 1997; Aberbach & Christensen, 2001; Hughes, 2003; Jones & Kettl, 2003). During the 1980s, many governments faced economic slowdowns and large public sectors that generated high levels of public expenditure[1]. What resulted was a "reassessment of the fundamental responsibilities of government in economic management and delivery of public service" (Kaul, 1997:14).

10.3 PUBLIC-PRIVATE PARTNERSHIPS

The above-mentioned reassessment of the government's role in providing services took place during the International Drinking Water Decade (1981–1990). This Decade had the ambitious target of ensuring adequate water services for all. Initially, the approach followed during this Decade adhered to the infrastructural approach sketched above. The emphasis was on constructing infrastructure and technically equipping a utility to provide services. Although services were expanded to millions, the Decade closed with rather disappointing results. One of the main lessons learnt was that a technological focus was not enough. The new credo was that "institutions matter."

In this context, private sector participation in the delivery of water services became increasingly advocated by a coalition of international development banks, bi-lateral donors and water sector professionals. International aid agencies saw the private sector as "an opportunity to deliver similar reforms through private sector approaches in lower economies" (Franceys, 2008:46). The private sector was viewed as more efficient and effective. The World Development Report *Infrastructure for Development* (World Bank, 1994), for instance, compared the large investments in infrastructures with their limited impact on development, and proposed private participation as a means to improve efficiency. The private sector was considered a viable option also for the poor areas, where services were most lacking (Idelovitch & Ringskog, 1995).

At the turn of the 21st century, private sector participation had reached its acme. Despite this, private operators always remained a minor provider in developing countries. In 1997 at least 93 countries had engaged in some form of private participation in the sector. This, however, mostly occurred in developed regions, while in developing countries involvement remained minor. At its peak, private sector participation was serving 14.4% of the urban population in middle and low income countries (Franceys, 2008).

Table 10.2 Regional trends in water and sanitation privatization (1997).

Country or region	Percentage of water and sanitation service delivery privatized 1997
Western Europe	20
United States	14
Central/Eastern Europe	5
Latin America	4
Africa	3
Asia	1

Source: Hall (1999) and National Resources Council (2004), as cited in Davis (2005).

[1]Hughes (2003:49) mentions the first oil shock as severely constraining government resources. As "practical politics" dictated that the actual services delivered by the public sector could not be cut, a strong pressure existed to attempt to manage the same services with less money and fewer staff.

10.4 FROM PRIVATIZATION TO PUBLIC-PRIVATE PARTNERSHIPS

The term "public-private partnership" entered the mainstream of the water services sector in the early 1990s. Over time it acquired two main meanings. On the one hand, some authors view "partnership" as a "governance tool", which transcends the government. To them a public-private partnership is a means of combining the "best of two worlds," the private and the public sphere (Brinkerhoff, 2002). On the other hand, others have argued that the term is no more than a "game of language" (Hodge & Greve, 2007). To them the purpose of using the term is to disguise privatization and outsourcing by "packaging" them into the more acceptable concept of partnerships. In this view, public-private partnerships are just "another chapter in the history of privatization" (Hodge & Greve, 2007:548). Franceys (2008:46) finds that these two dimensions can be reconciled and the choice of the term "served the dual purpose of appearing more acceptable to society than "privatization", but is also a necessary description of how government, consumer and providers, mediated by some form of economic regulation, necessarily have to recognize each others' needs and interests".

Public-private partnerships were advocated with the promise of increased efficiency, reduced pressure on public budgets by tapping into private sector investment funds, and increased technical and managerial expertise (Hodge & Greve, 2007; Brown, 2002; Idelovitch & Ringskog, 1995; World Bank, 1994). Moreover, these partnerships were perceived as a strategy to disconnect politics from the management of water services (Brown, 2002). The partnerships themselves concerned a wide variety of contractual arrangements ranging from service contracts to concession contracts[2]. The role of the government varies depending on the type of arrangement (Figure 10.1).

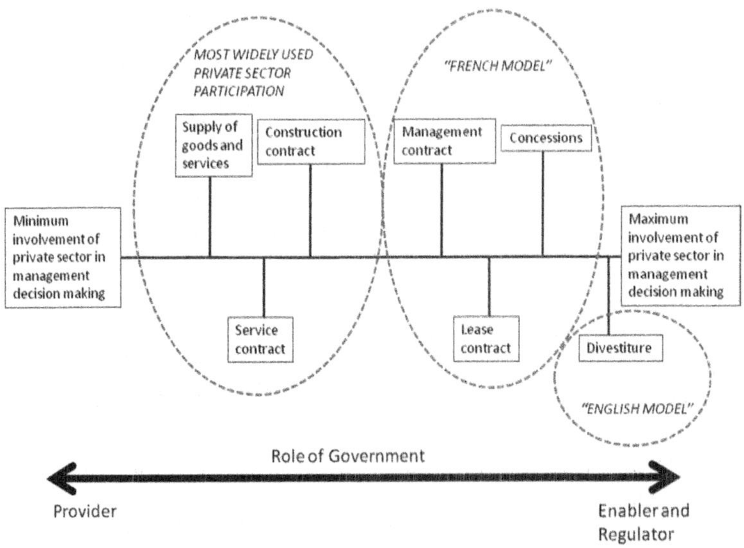

Figure 10.1 Public-Private Partnerships and the changing role of government.
Source: Pietilä, 2012; personal communication.

At the core of these partnerships is essentially the replacement of a public service provider by a private one. Whereas the twinning projects focused on capacitating one of the partnering utilities, public-private

[2]For a description of the different types of contractual arrangements see World Bank (1997).

partnerships essentially sought to replace the public provider by a private provider. The underlying infrastructural focus (the modern infrastructural ideal) remains the same (universal provision of a standardized service by a single provider). The only difference is that a public monopoly is replaced by a (regulated) private monopoly.

Despite heavy promotion from international financing agencies and donors, public-private partnerships have remained controversial in the water services sector (Hall, 2000; Barlow & Clarke, 2003). Interestingly, the criticism does not focus on the partnership itself but rather on private sector involvement and the pros and cons of private sector participation. Studies comparing the performance of private and public utilities generally come to the conclusion that neither is more efficient than the other (Donahue, 1989; Braadbaart, 2002; Prasad, 2006). The anticipated injection of private sector funds has also remained limited. Despite some findings that public-private partnerships have led to an acceleration of and increase in capital investments (Davis, 2005), in most cases these investments were not made by the private partner. Actually, development aid agencies and International Banks made the capital available *because* of the involvement of a private partner (Franceys, 2008). Moreover, when the private operators did invest, it was usually in developed countries (Table 10.3).

Table 10.3 Levels of investment and percentage of investments in Africa in 2004.

Company	Total investment in the water sector	Percentage of those investment in Africa
SAUR	€ 22 Billion	3.7%
VEOLIA	€ 46 Billion	Less than 4%
SUEZ	€ 42 Billion	Less than 4%

Source: Talbot, 2004, as cited in Schwartz, 2008.

The envisaged separation of politics from service provision has also remained elusive. Brown (2002:126), in arguing why privatization does not lead to a separation of politics from service provision, provides the following observation, which has largely held true for the water services sector:

"*It is absurd to think that a private [operator] will sink a large amount of capital into an enterprise and then unilaterally disarm himself politically. Obviously investors will use all legal means, including political, to protect their interests. Similarly, it is unreal to expect that social expectations will terminate or diminish merely because [responsibilities] have been transferred to the [private sector]. It seems obvious that investors will seek to manipulate a system to their benefit and equally obvious that politicians, interest groups, advocacy organizations and others will continue to push for their own objectives*".

Moreover, by the beginning of the 21st century large international operators increasingly became unwilling to engage in the water services sectors of developing countries. Rather, they opted for concentrating on more "sound markets in Europe and North America (SUEZ, 2003).

10.5 WATER OPERATOR PARTNERSHIPS

As the relative "popularity" of Public-Private Partnerships decreased, the concept of "Water Operator Partnerships"[3] (WOPs) surfaced as a possible alternative for improving service provision in developing

[3]See IWA *et al.* (2009) for a more detailed explanation of water operator partnerships; see also Patron Coppel and Schwartz, (2011) for a case study of WOPs in Mozambique.

countries. In March 2006, the United Nations Secretary General Advisory Board in Water and Sanitation (UNSGAB) led by the ex-prime minister of Japan, Ryutaro Hashimoto, developed a Compendium of Actions titled the Hashimoto Action Plan (Hashimoto, 2006) under the direct request of the Secretary General Kofi Annan. The objective of this Plan was to put forward a series of actions aimed at boosting the efforts around the world for achieving the MDG targets on water and sanitation. The UNSGAB recommended WOPs as a useful mechanism for that.

The concept of WOPs has since been heavily promoted by organizations such as the International Water Association, UN-HABITAT, and individual utilities seeking to internationalize their activities. However, there still appears to be no generally agreed upon definition for what a WOP exactly is. The International Water Association (IWA) defines WOPs as "any formal or informal collaboration or structured partnership aimed at capacity building on a not-for-profit basis. Partnerships can take a multitude of different forms and have various technical, legal and social shapes depending on individual circumstances" (IWA, *et al.* 2009).

Like twinning projects and public-private partnerships, WOPs also tend to adhere to the modern infrastructural ideal. WOPs, however, do represent a marked break from the public-private partnership approach. A fundamental element of the public-private partnership approach was essentially the replacement of a public organization or utility by a private organization. In the WOP approach, the emphasis is on capacitating (rather than replacing) the public organization. At the same time, WOPs do not represent a return to twinning projects. The main difference is that WOPs tend to incorporate private sector institutions (emphasis on cost-recovery, efficiency, performance-based management, etc.). Illustrative of this approach is the Water Operator Partnership developed by Vitens-Evides International in Mozambique. Underlying this partnership is a project document, which mimics a management contract as a way to set targets (performance indicators) for the partnership. It was felt that this "contract" would help formalize the project and give it some structure.

Underlying these water operator partnerships are two, rather contradictory, claims. First, the projects are undertaken on a so-called "not-for-profit, not-for-loss"[4] basis. Essentially the partners involved in the project will not lose any money on these projects. At the same time, these projects claim to be based on solidarity between utilities. As the partners do not lose money on these projects, the question is what is the basis of solidarity in these projects? Given that the funds for these projects either come from donors, international development agencies or tariffs imposed on customers of utilities in developed countries, it would seem that the solidarity stems more from the customers or taxpayers than the public utility.

Despite the promotion of WOPs by organizations like IWA and UN-HABITAT, the actual number of projects developed under this umbrella is rather limited. The main limitations appear to be two-fold. On the one hand, running a utility in a developed country is very different from operating a service provider beset by problems such as servicing informal settlements, high levels of non-revenue water and illegal connections. Developing these projects thus entails developing a new kind of capacity in the utility. Few utilities seem to be willing to do this. Secondly, partnerships of any kind will only work if they are linked to investments in infrastructure. Funds available for such investments are limited.

[4]The concept of "not-for-profit" projects is rather problematic. Only once the project has ended, can the extent to which the project was not-for-profit or indeed profitable be determined. Moreover, these projects often serve secondary purposes, such as making the company attractive for new staff, improving the corporate image of the utility and so on. Achieving these benefits, which normally would cost money, is now essentially subsidized.

10.6 PARTNERSHIPS WITH SMALL-SCALE PROVIDERS

Although small-scale independent providers (SSIPs) have a long history of filling the service provision gap left by "formal" utilities, these providers have been largely ignored by international lending agencies, donors and national governments (Collignon & Vézina, 2000). Since 2003, however, partnerships involving a water utility partnering with small-scale independent providers (SSIPs) who undertake part of the distribution of the service provision process, have gained prominence. This partnership differs from those described previously by diverging from the modern infrastructural ideal and placing its emphasis on the single service provider. This partnership thus entails multiple service providers, who may also be providing services of varying levels and at different prices. Important to note is that these small-scale providers are not a replacement of the formal utility (which most likely was not providing services to the clientele of the SSIPs), but rather complement service provision of the utility.

Two reasons appear to underlie the increasing importance of small-scale providers. First of all, the challenge of achieving the MDGs targets for water supply and sanitation has led to a reconsideration of existing modes of service delivery. The MDG targets for water supply services state that the population without access to adequate water services should be halved by 2015. Although progress is reported on these targets, especially access to adequate water services in Sub-Saharan Africa is lacking. Between 1990 and 2006 the percentage of people with access to adequate water services increased from 49% to 58%, which represents an increase of only 9% in 16 years. Considering that almost 75% coverage should be achieved by 2015, considerable (additional) efforts are needed to achieve the MDGs (WHO/UNICEF, 2008). By formalizing SSIPs and bringing them into the formal service provision system, considerable gains can be made in terms of achieving the MDGs (see Eales, 2008). Secondly, following the disappointing results of the "water privatization decade" which ended in 2003 (Franceys, 2008:45), the proponents of private sector involvement have been searching for new private sector champions in the water services sector. One of the ways in which the role of the private sector is again receiving attention is through an increased focus on small-scale independent providers. Increasingly, the entrepreneurial spirit of SSIPs is being emphasized by describing them as "small-scale entrepreneurs" (Eales, 2008), "independent water entrepreneurs" (Solo, 2003) and "local entrepreneurs" (Conan & Paniagua, 2003).

10.7 CONCLUSION

In this chapter we have presented different types of partnerships and tried to link them to different time periods and different approaches to the water service sector. The different partnership types represent different perspectives in terms of the objectives to be achieved by the partnership, the nature of the partners involved in such a partnership, the guiding principles under which such a partnership takes place and, the level of service that is provided. Underlying these different approaches are different problem definitions for challenges facing the sector. These problem definitions have changed over time and the types of partnerships promoted to address these challenges have also changed.

The initial twinning project had a strong infrastructure focus and the challenges facing the sector were considered mainly to be technical in nature. The challenges could be addressed by increasing technical capacity and developing more infrastructure. The technical focus is largely discredited now at the end of the Drinking Water Decade. Institutions do matter which in the prevailing neo-liberal climate quickly translates into increased need for private sector participation. Public-Private Partnerships were the vehicle to achieve this increase in private sector participation. At the beginning of the 21st century, however, the drive towards increased private sector participation has lost momentum. Under pressure to achieve the MDGs, alternative forms of partnerships are sought. The Water Operator Partnership, with its claims of solidarity and capacity development, and partnerships involving small-scale providers, with their emphasis on "small-scale entrepreneurs" appear to be the most advocated alternative forms. The success of these forms will become evident in coming years.

REFERENCES

Aberbach J. and Christensen T. (2001). Radical reform in New Zealand: crisis, windows of opportunity, and rational actors. *Public Administration*, **79**(2), 403–422.

Barlow M. and Clarke T. (2003). Blue Gold: The Battle of Corporate Theft of the World's Water. Earthscan Publications Limited, London.

Barzelay M. (2001). The New Public Management: Improving Research and Policy Dialogue. University of California Press, Berkeley.

Batley R. (2006). Engaged or divorced? Cross-service findings on government relations with non-state service-providers. *Public Administration and Development*, **26**, 241–251.

Boag G. and McDonald D. A. (2010). A critical review of public-public partnerships in water services. *Water Alternatives*, **3**(1), 1–25.

Braadbaart O. (2002). Private versus public provision of water services: does ownership matter for utility efficiency? *Journal of Water Supply: Research and Technology – AQUA*, **51**(7), 375–388.

Brikenhoff J. (2002). Assessing and improving partnership relationships and outcomes: A proposed framework. *Evaluation and Program Planning*, **25**, 215–231.

Brikenhoff J. and Brinkerhoff D. (2002). Government-non profit partnership: a defining framework. *Public Administration and Development*, **22**, 19–30.

Brown A. (2002). 'Confusing means and ends: framework of restructuring, not privatization, matters most'. *International Journal of Regulation and Governance*, **1**(2), 115–128.

Chaplin S. E. (1999). Cities, sewers and poverty: India's politics of sanitation. *Environment and Urbanization*, **11**, 145–158.

Collignon B. and Vézina M. (2000). Independent Water and Sanitation Providers in African Cities. Full Report of a Ten-Country Study. Water and Sanitation Program, Washington, DC.

Conan H. and Paniagua M. (2003). The Roll of Small Scale Private Water Providers in Serving the Poor. Asian Development Bank, Manila.

Donahue J. (1989). The Privatization Decision: Public Ends, Private Means. Basic Books, New York.

Eales K. (2008). Partnerships for sanitation for the urban poor: is it time to shift paradigm? IRC Symposium on Sanitation for the Urban Poor. IRC International Water and Sanitation Centre, The Hague.

Franceys R. (2008). GATS, 'Privatization' and institutional development for urban water provision: Future postponed? *Progress in Development Studies*, **8**(1), 45–58.

Hall D. and Lobina E. (2000). Public sector alternatives to water supply and sewerage privatization: case studies. *International Journal of Water Resources Development*, **16**(1), 35–55.

Hashimoto R. (2006). Hashimoto Action Plan - Compendium of Action. United Nations Secretary-General's Advisory Board on Water and Sanitation, New York.

Hauck V. and Land T. (2000). Beyond the Partnership Rhetoric: Reviewing Experiences and Policy Considerations for Implementing 'Genuine' Partnerships in North-South Cooperation. ECDPM Discussion Paper 20, Maastricht: ECDPM.

Hodge G. A. and Greve C. (2007). Public–private partnerships: an international performance review. *Public Administration Review*, **67**(3), 545–558.

Hood C. (1991). A public management for all seasons. *Public Administration*, **69**, 3–19.

Hughes O. (2003). Public Management and Administration: An Introduction. Third edition, Palgrave MacMillan, New York.

Idelovitch E. and Ringskog K. (1995). Private Sector Participation in Water Supply and Sanitation in Latin America. World Bank, Washington, DC.

IWA, GWOPA and VEI. (2009). Water Operators Partnerships - Building WOPs for Sustainable Development in Water and Sanitation. IWA, London.

Jones L. and Kettl D. (2003). Assessing public management reform in an international context. *International Public Management Review*, **4**(1), 1–19.

Kaul M. (1997). The new public administration: management innovations in government. *Public Administration and Development*, **17**, 13–26.

Lariola M., Öhlund S., Håkansson B. and Emsis I. (2000). Twinning Cooperation between Riga Water Company and Stockholm Water Company. SIDA Evaluation 00/7. Department for Central and Eastern Europe. SIDA, Stockholm.

Lister S. (2000). Power in partnership? An analysis of an NGO's relationships with its partners. *Journal of International Development*, **12**, 227–239.

Minogue M. (1997). Theory and practice in public policy and administration. In: The Policy Process: A Reader, M. Hill (ed.), second edition, Pearson Education Limited, Harlow.

Nilsson D. (2006). A heritage of unsustainability? Reviewing the origin of the large-scale water and sanitation system in Kampala, Uganda. *Environment and Urbanization*, **18**, 369–385.

Patron Coppel G. and Schwartz K. (2011). Water Operator Partnerships as a model to achieve the MDGs on water supply? Lessons from four cities in Mozambique. *Water SA*, **37**(4), 575–584.

Prasad N. (2006). Privatisation results: private sector participation in water services after 15 years. *Development Policy Review*, **24**(6), 669–692.

Schwartz K. (2008). The new public management: the future for reforms in the African water supply and sanitation sector? *Utilities Policy*, **16**(1), 49–58.

Solo T. 1999 Small-scale entrepreneurs in the urban water and sanitation market. *Environment and Urbanization*, **11**(1), 117–131.

SUEZ (2003). SUEZ Introduces its 2003-2004 Action Plan: Refocus, Reduce Debt, Increase Profitability. Press release. SUEZ, Paris.

WHO/UNICEF (2008). Progress on Drinking Water and Sanitation: Special Focus on Sanitation. UNICEF and Geneva: WHO, New York.

World Bank (1994). World Development Report 1994: Infrastructure for Development. World Development Report. World Bank, Washington, DC.

World Bank (1997). Toolkits for Private Participation in Water and Sanitation. World Bank, Washington, DC.

Chapter 11

The birth, growth and decline of multinational water companies

David Hall and Emanuele Lobina

11.1 PRIVATE COMPANIES IN HISTORICAL CONTEXT

Private watercompanies were created in the 19th century as industrialisation and urbanisation took off. By the early 20th century, these had been replaced almost everywhere by municipal water companies, as the private companies lacked the capacity for large-scale investment and required higher rates of return on selected market segments. Only in France, and to a much smaller extent in Spain, did private companies of any size survive the wave of municipalisation of water systems which took place elsewhere in Europe and the USA. The survival of these companies, and the size of the market, depended crucially on negotiations with cities over the political decisions to municipalise services or not. (Juuti & Katko, 2005).

The two companies which still dominate private water were both created in the 19th century. Generale des Eaux, now part of Veolia, was created in 1853 by charter of the emperor Napoleon III; Lyonnaise des Eaux, now the core of Suez Environnement, was created in 1880. In France itself, the companies not only retained their positions during the 20th century but extended them. As piped water networks were extended, especially after World War II, the companies obtained an increasing proportion of the sector through lease ("affermage") contracts with municipalities. By the end of the 20th century, over 70% of French water services were run by private companies. (AFD, 2010)[1].

The subsequent international growth of the French companies was also assisted by the international expansion of France, and French capital, in the 19th and 20th centuries. The first international contracts were gained in the 1880s by Generale des Eaux, including four concessions in Italy-Verona, Naples, Bergamo and Venice; Porto (Portugal) and, in 1882, Constantinople (now Istanbul), which was then the capital of the Ottoman Empire. Contracts and concessions were also obtained in French colonies, with Lyonnaise des Eaux active in Algéria, Morocco, Tunisia, Madagascar, Guinéa, Congo, Central African republic, Sénégal, and Togo. Contracts in Cote d'Ivoire contributed a lot to the growth of the third, much smaller French water company, SAUR. (AFD, 2010)[2].

[1]AFD, 2010 Water services and the private sectorin developing countries http://64.95.129.92/ppiaf/sites/ppiaf.org/files/publication/water-services-and-the-private-sector-in-developing-countries.pdf#page=30.

[2]AFD, 2010 Water services and the private sectorin developing countries http://64.95.129.92/ppiaf/sites/ppiaf.org/files/publication/water-services-and-the-private-sector-in-developing-countries.pdf#page=30.

The French companies were also able to buy dominant positions in the residual private water sector in other countries, both extending their own business and eliminating potential competitors. The biggest of the private Spanish water companies, Aguas de Barcelona, which originated as a French venture, was by the 1950s again effectively controlled by Suez; another, Aguas de Valencia, was controlled by SAUR. Others were mostly owned by Spanish construction groups, such as FCC (which was itself controlled by Veolia for a few years in the early 2000s).

Competition has played a very minor role in these developments. The private business in this sector has been built through politically negotiated monopolies, which were then retained for decades, in some cases over a century. In France, at the end of 2008, over 95% of private contracts in the towns and cities with over 100,000 population were held by the company which was originally awarded it without competitive tendering procedures, and no non-French company has ever won a single contract.[3] The established oligopoly of the three French companies seeks to maintain its position. In 2002 they were the subject of a French competition commission report for colluding through joint ventures, and in 2012 they were being investigated for collusion by the European Commissions. In Spain, too, two-thirds of private contracts have never been subject to competition, and three cities – Barcelona, Valencia, Alicante – have private water operations which have run for over 100 years without any competition. The case of Barcelona is remarkable because there is in fact no legal concession contract, according to a recent court ruling.[4] (Hall & Lobina, 2012).

11.2 RISE AND FALL

This pattern remained broadly unchanged for most of the 20th century, until the sudden expansion of private water companies from the late 1980s. It was also made possible by political developments, both at national and international level.

The biggest national change came in the UK in 1989, when the Thatcher government privatised water throughout England and Wales, against strong public opposition, by floating the 10 regional companies on the stock exchange. This was only possible because England and Wales, uniquely in Western Europe, had restructured their water sector 15 years earlier, so that all municipal operations had been merged into a small number of state-owned regional companies. (Hall & Lobina, 2007).

This had a number of consequences. It gave almost the entire sector in a large European country to private companies, thus nearly doubling the private share of the market at a stroke. It created a new set of companies with very secure finances, which might be potential competitors to the French groups in other countries, although the UK companies were at a relative disadvantage because of their total lack of experience in negotiating with public authorities to gain new contracts. And it established water privatisation as a political possibility which could be emulated globally.

Even after this, any further growth of private companies had to take place across two dimensions. Firstly, expansion had to involve international growth into other countries, because the home markets of the private companies (France, England & Wales) were effectively saturated and already controlled by a small group of companies under conditions of secure monopoly. Secondly, because the public sector is so dominant, there were few existing pockets of private business to be acquired, so obtaining new business required reducing the role of the public sector in any given country. The companies not only had to conduct a strategy of internationalisation, but also a strategy of reducing the role of the public sector.

---

[3]Observatoire des DSP: Déroulement des procédures de délégation des services publics d'eau et d'assainissement, procédures 2004. http://www.agroparistech.fr/labogea/doc_dsp.html : "Aucune entreprise étrangère n'a encore pénétré le marché".

[4]JutJatContenciósAdministratiu N° 12 De Barcelona Recurs Núm.: 376/2008 2c Sentencia núm. 298/2010.

The political conditions for both these elements were created by the international financial institutions – themselves, ironically, public sector institutions. The World Bank, the IMF and other development banks adopted neoliberal principles that development depended on maximising the role of the private sector. Publications and conferences promoted the virtues of privatisation, especially in infrastructure sectors such as water, as a way of delivering both investment and efficiency in these sectors in developing countries. The banks and the IMFalso started including privatisation as a condition of their loans in the sector. If a country wanted such loans, it would have to privatise. The loans themselves were also vital to the companies, as they provided large amounts of investment finance at low interest rates. (Hall & Lobina, 2007).

The promotional activity and the conditionalities launched a wave of water privatisations. The flagship in the south was the privatisation of water in Buenos Aires in 1992, when the Argentinian government was the "model pupil" of the IMF. There was a competitive tender, but the winning consortium, Aguas Argentinas, included not only Suez but also Veolia, Aguas de Barcelona, and Anglian Water, which considerably reduced potential opposition. Other contracts in Argentina rapidly followed, along with a concession in Cartagena in Colombia, Guayaquil in Ecuador, and a few in Brazil. There was significant resistance in Brazil, however, and attempts to privatise water in Rio de Janeiro and some other cities were halted. At the end of the 1990s, Chile started privatising its water operations. But from 2000, the bandwagon ran into major economic and political problems in the region. In Bolivia, there was a massive uprising in the city of Cochabamba, and a similar uprising later against the privatisation of water in La Paz and its neighbouring city, El Alto. In 2000, the Argentinian economy collapsed, and the companies were unable to persuade the new political leaders to continue the concessions or to compensate the companies for their losses. In Uruguay, a referendum agreed to a constitutional amendment declaring water to be a human right and making privatisation illegal. (Lobina & Hall, 2007).

Similar developments took place in Asia and Africa, though more selectively. The city of Jakarta in Indonesia was divided into two contracts, one for Thames Water and one for Suez, both in partnership with companies owned by cronies of the dictator Suharto. After Suharto's fall, the companies clung onto these contracts, which include guaranteed levels of profit. In the Philippines, Manila was also divided, this time between Suez and United Utilities. The Asian financial crisis upset the economics of these concessions, which required major renegotiation and debt relief from the state, but the result has been profitable business. There were very few concessions elsewhere in Asia until China was identified as a major future market by both Suez and Veolia, who both established joint ventures with Chinese companies. They have obtained a few contracts, and remain optimistic, if only because even a tiny share of the market in China is still significant. (Hall & Lobina, 2006).

In Africa, new contracts were added to the existing private water deals in former French colonies. Suez obtained contracts through a joint venture in South Africa, still under the apartheid regime, and later a management contract in Johannesburg. But the political opposition prevented any major inroads into the public sector, with the exception of two concessions run by Biwater. Elsewhere in sub-Saharan Africa a number of contracts were awarded, invariably with strong support from the World Bank, and without any commitment to investment, but by the mid-2000s many of these were not economically viable. In North Africa, the major expansion took place in Morocco, where the business-friendly dictatorship of King Hassan II allowed both Suez and Veolia to win major concessions. But during the Arab spring of 2011, these concessions were the focus of some angry demonstrations, followed by scathing public audit reports, which have resulted in a write-down of €59 million by Veolia. (Hall & Lobina, 2006, 2012).

Significant expansion also took place in Central and Eastern Europe, where the collapse of communism opened up large new potential business in sectors that had previously been part of the state. The process started in 1992, with the concession obtained in Gdansk (Poland) by SAUR, and continued with a rapid expansion of contracts in most of the main towns and cities of the Czech Republic and Hungary by the end of the decade, together with some privatisations in Romania, Bulgaria (Sofia), and Estonia (Tallinn). The French companies dominated this process. The UK companies did participate in the Czech Republic, where the private stake in SCVK, for example, was originally bought by Hyder (the parent group of Welsh Water in the late 1990s), and Anglian Water, acquired stakes in three companies. But even in Eastern Europe there was strong resistance. In Poland, for example, despite much publicity for the original deal in Gdansk, no other major city agreed to privatise its water. Few new contracts were won after the 1990s, and attempted privatisations in the former Soviet Union also failed to endure, as contracts were terminated in Tblisi, Georgia; Almaty, Kazakhstan; Odessa and Kirovograd, Ukraine; and Bukhara and Samarkand, Uzbekistan. By 2010 criticism of existing private water contracts was growing, and the Hungarian towns of Pecs and Kaposvar terminated contracts held by Suez in in 2009, and in 2012 the capital city of Budapest started the same process. There have also been strong pressures against the privatised contracts of United Utilities, which sold its business in Sofia to Veolia, but was unable to sell the operation in Tallinn, which has been the subject of criticism by the city council, consumers, the ombudsman and the national government. (Hall & Lobina, 2007, 2012).

The attempts to expand into other west European countries had little success except in the two countries where the French multinationals were already established: Spain and Italy. In Italy, a favourable environment was provided by the 1994 Galli Law which reformed the whole water sector, and by the partial privatisation of major utilities such as Acea in Rome, in which Suez bought a significant stake. Some new concessions were also acquired, for example in Arezzo. In Spain, Veolia built a dominant shareholding in FCC, the second largest private water operator in Spain, and used it as a partner in Latin America – as Suez did with Aguas de Barcelona. But attempts to expand in other countries within the EU proved unsuccessful, provoking considerable public and political opposition in nearly all countries. In Germany, for example, private companies were only able to win a few concessions in the former East German towns such as Rostock and Potsdam, apart from Berlin itself, where a fiscal crisis forced the partial sale of Berliner Wasser Betriebe. But the privatisation of Potsdam was terminated, and by 2011 there were strong demands for the Berlin privatisation to be undone.

Private companies have also found their existing markets being eroded by a new wave of re-municipalisations, most dramatically in France itself. This was led by the city of Paris itself, which decided to re-municipalise its water services from January 2010, after the two 25-year concessions which had been given to Suez and Veolia expired. The decision was driven both by the political view of the ruling socialist-green coalition that vital public services should be in public hands, and by an economic assessment that a publicly run service would be cheaper. This was demonstrated in the first year, when the city saved €35 million and was able to reduce water prices by 8%. Veolia managed to save its contract in the even larger Ile-de-France region around Paris, but by 2012 a further 40 French municipalities had also decided to follow the example of Paris and re-municipalise water services, including major cities such as Bordeaux and Brest[5].

Private companies were dealt a further blow by the Italian national referendum in 2011 which rejected proposals for liberalisation and privatisation of water services. This has also prevented the Italian government from selling water services as part of the privatisation programme required under the EU rescue deal. In other European countries subject to EU and IMF conditions, however, privatisation proposals include some public water companies, most notably in Greece, Portugal and Spain, where a new proposal to privatise the public water service of Madrid was put forward in 2011. (Hall & Lobina, 2012).

[5]Pigeon M. *et al.*, 2012 Remunicipalisation: putting water back into public hands. TNI http://www.corporateeurope.org/publications/remunicipalisation-putting-water-back-public-hands.

The expansion of the 1990s also involved consolidation of existing pockets of privatised water. All three French companies made acquisitions in the UK, the largest being Suez' acquisition of Northumbrian Water, one of the newly privatised regional monopolies, while both Veolia and SAUR built substantial holdings in the older and smaller water-only companies. Both the French and the British companies bought private companies in the USA, where up to 15% of the services had remained in private hands. But during the 2000s the companies began to lose contracts in the USA as well, for example in Atlanta in 2003, and by the end of the decade public opposition to privatised water was as strong in the USA as elsewhere. There were also concessions obtained in Australia, where Thames and Veolia partnered a joint venture in Adelaide, and some smaller concessions obtained by Anglian water in New Zealand.

During the 2000s, the English followed by French companies decided that most international expansion should be abandoned. They recognised the political reality that some existing contracts were being terminated anyway, and the economic reality that they were unable to deliver reliable returns. In 2003 Suez announced it would withdraw a large proportion of its contracts, and not take on any new contracts without a secure, high level of return. Veolia's withdrawal has been more erratic, but by 2011 it announced it would leave nearly half of the 77 countries in which it was doing business. The British companies withdrew almost completely, except for the privately owned Biwater, which continued to operate global water contracts until it sold them to Sembcorp in 2010. The private equity firms which took over many British companies were especially insistent on ending the international ventures: for example, Maquarie insisted that all of Thames Water's overseas operations should be sold before it would complete the takeover in 2006. The same was apparent in France: when Bouygues sold SAUR to a private equity firm, it had to retain most of the overseas contracts.

Internationally, private companies have become increasingly dependent on the support and partnership of development banks as equity shareholders in their international activity. The International Finance Corporation (IFC) invested $25 million in 13.9% of the shares of Veolia AMI, the company's subsidiary aimed at Africa, Middle East and India; Proparco, an arm of the French state's Agence Française de Développement (AFD), invested 10 million in buying 5.6% of the shares.[6] The IFC also invested €100 million in buying 9.5% of the shares of Veolia Voda, Veolia's Eastern European arm, in which the EBRD has also invested €175 million.[7] The EBRD has provided nearly €500 million in equity and loan finance for private water companies since 1991 (Table 11.1).

Table 11.1 EBRD finance for private water, 1991–2009 € million.

	EBRD finance 1991–2009	Of which equity investments
Suez	42	0
Veolia	263	175
FCC/Aqualia	80	80
United Utilities	111	17
TOTAL	*496*	*272*
Veolia other (heating, transport)	*208*	*141*

Source: EBRD investments 1991–2009 http://www.ebrd.com/downloads/research/annual/invest09.xls.

[6]Veolia press release, January 09, 2008. The International Financial Corporation (IFC) and PROPARCO invest in Veolia Water Africa, Middle East and India.

[7]Veolia http://www.finance.veolia.com/docs/Operating-and-Financial-Review-June-2010-def-sans-mark-up.pdf; IFC http://www.ifc.org/ifcext/media.nsf/content/SelectedPressRelease?OpenDocument&UNID=80FBA1D1C466866085257737005B9ABF; EBRD investments 1991–2009 http://www.ebrd.com/downloads/research/annual/invest09.xls.

The rise and fall are dramatic in policy terms, and in terms of company balance sheets. But in historical context it represents only a small deviation from the long-term dominance of the public sector model of water supply services. At the peak of the expansion, around 2000, the proportion of the world's largest cities with private water supply was never more than about 11%, and by 2012 it was declining to about 9%. Outside these cities, the proportions are certainly much smaller. In the 21st century, as in the 20th century, water remains a public sector service.

11.3 THE COMPANIES IN 2012

11.3.1 French companies: depending on the state

The final irony is the partial nationalisation of the private French companies themselves. As they weakened financially in the late 2000s, all the companies became vulnerable to takeovers, and a national strategy to protect key French companies from foreign control resulted in the French state becoming the major shareholder in all three. In April 2007 SAUR was bought by a consortium led by the French state, which still holds 38%, in order to prevent a foreign private equity takeover.[8] In 2009 Suez was merged with the state-owned GdF, with the result that Suez Environnement, including all the water operations, is now 36% owned by GdF-Suez, which is itself 36% owned by the French state. The impact of this extends beyond France, since Aguas de Barcelona is now almost wholly owned by Suez. The French state has also become the largest shareholder in Veolia, with 12.4% of the shares, but its energy division Dalkia is already a 50–50 venture with EdF, which is 85% state-owned. A similar attempt to merge its transport division into a 50–50 venture with a state-owned company failed, and the French state is now effectively taking responsibility for the loss-making transport operations. Faced with declining returns and large debts, and exacerbated by the recession, Veolia wrote off over €800 million in 2011, and set itself targets of closing operations in nearly half of the 77 countries in which it operates and selling €5 billion worth of assets. (Hall & Lobina, 2012; Veolia, 2012; Suez, 2012).

11.3.2 English companies taken over

The withdrawal of English companies from international activity was almost completed in 2010, when Biwater and United Utilities sold their international operations – with the exception of United Utilities, which was unable to sell its shares in Tallinn. These companies have not only withdrawn from international activity, but have themselves become the subject of takeovers. Of the 10 large water and sewerage companies, four – Anglian, Southern, Thames and Yorkshire – are already owned by private equity or financial groups. Three large companies are still part of groups quoted on the London stock exchange – Severn Trent, South West, and United Utilities. Two of the 10 large water and sewerage companies are owned by Asian multinationals – Wessex by the Malaysian company YTL, and Northumbrian was bought in 2011 by the Hong Kong group Cheung Kong Infrastructure. The remaining one, Welsh Water, is owned by a not-for-profit private company (Glas Cymru). Of the smaller water-only companies, only one is still listed on the stock exchange, Dee Valley, and it is 35% owned by insurance company Axa. Most of the others are owned by various private equity funds, and the French groups are finally selling their holdings in these companies: Suez has sold 70% of Bristol Water to a Canadian infrastructure company, and Veolia is expected to sell its subsidiaries to a group of private equity and

[8]Les Echos April 20, 2007 Friday : FRENCH CONSORTIUM ACQUIRES SAUR (LA SAUR REPRISE PAR LE CONSORTIUM COMPOSE DE LA CDC, SECHE ET AXA IM.

financial investors in 2012. The remaining company, Bournemouth, is owned by the Singapore group Sembcorp as a result of its takeover of Biwater. (Hall & Lobina, 2012) (Table 11.2).

Table 11.2 England and wales: Water company ownership, april 2012.

Company	Owner	Country	Type of owner	Comments
Anglian Water	Osprey/AWG	UK	PE	Consortium of 3 PE funds, inc. 3i
Northumbrian Water	Cheung Kong Infr.	UK	M	via UK Water
Severn Trent Water	Severn Trent	UK	SEC	
Southern Water	Greensands	UK	PE	Consortium: IIF 28%, Challenger 23%, UBS 16%
South West Water	Pennon Group	UK	SEC	
Thames Water	Macquarie	Australia	PE	Also China investment Corp CIC 8.7%, Abu Dhabi Investment Corp 9.9%
United Utilities Water	United Utilities	UK	SEC	
Welsh Water	GlasCymru	UK	NPC	Not for profit private company
Wessex Water	YTL	Malaysia	M	Malaysian power company
Yorkshire Water	Saltaire Water	UK	PE	PE consortium: Citi, GIC, Infracapital
Bournemouth and West Hampshire Water	Sembcorp	Singapore	M	Singapore based company http://www.sembcorp.com/
Bristol Water	Capstone (70%), Agbar/Suez (30%)	Canada, ES/FR	PE,M	Listed infrastructure fund, spun off by Maquarie
Cambridge Water	Alinda	USA	PE	Alinda Capital partners, a US PE firm
Cholderton Water	Cholderton Estate	UK	P	Private family owned
Dee Valley	–	UK	SEC	35% of shares owned by Axa SA.
Folkestone and Dover	Veolia	FR	M	
Portsmouth Water	South Downs Capital	UK	PE	15% owned by directors, 36% owned by SMIF/Land Securities
South East Water	UTA and HDF	Australia	PE	Utilities Trust of Australia, Hastings Diversified Utilities Fund
South Staffordshire Water	Alinda	USA	PE	Alinda Capital partners, a US PE firm
Sutton & East Surrey Water	Aqueduct Capital	DE	PE	Canadian funds CDPQ 25%, AIM 17.5%, PSP 22%; plus Aqueduct Capital 25%, GBP 10%
Tendring Hundred	Veolia	FR	M	
Three Valleys	Veolia	FR	M	

Type of owner: SEC = stock exchange quoted (UK); M = multinational; PE = private equity; NPC = not-for-profit company; P = privately owned company.

11.3.3 Other companies

There were attempts by other companies, particularly energy companies, to enter the sector in the 1990s, but they failed. The US-based energy company Enron attempted to break into the market by acquiring the UK company Wessex Water and then forming an international company, Azurix. This proved a failure, with Enron deciding to break up Azurix and sell its assets even before Enron's accounting brought about its own collapse. One of the main reasons for Azurix' failure was the poor results obtained when bidding against Veolia and Suez: Azurix at one stage announced that it would only focus on smaller projects which would not appeal to the French groups. Despite frequent publicity, US companies such as CH2M Hill never expanded internationally into the water supply business, as opposed to engineering and consultancy. (Hall & Lobina, 2007).

German energy conglomerate RWE also expanded into the water sector in the late 1990s, both in Germany and beyond. Rather than competing directly, it established a number of joint ventures with Suez and Veolia, including the acquisition of a 49.9% stake in Berlin's water company Berliner Wasser Betriebe jointly with Veolia; an indirect participation in Budapest Municipal Sewerage Company FCSM through Berliner Wasser Betriebe, with Veolia as the other partner; and the acquisition of a 25% stake in Budapest Waterworks Rt jointly with Suez-Lyonnaise des Eaux. In 2000 it bought the UK company, Thames Water, which had already acquired a significant share of the world water market, but in 2006 RWE sold Thames to a private equity company, having failed to establish a profitable international presence outside the UK. The other major German energy group, EON, bought the German water company Gelsenwasser and made some efforts to expand, mainly in central and eastern Europe and the Mediterranean, but later sold it to German municipalities: it still holds a few contracts in Hungary and Poland. An Austrian energy utility, Energie AG, similarly acquired a few contracts in other Central European countries.

The World Bank and the OECD have argued that there is a "new breed" of water companies, including multinationals, based in southern countries. A PPIAF paper claimed that "It would be hard to overestimate the importance of this new trend", but their list of such companies is unconvincing. It includes a number of companies which no longer own or operate the water services listed, one 100% public sector company, some historical oddities, and companies owned by private equity firms based in tax havens. (Marin, 2009).

It includes two of the groups which have bought water companies in the UK, YTL and Cheung Kong, but these groups so far have no other interests in the water sector, and the UK investments may more accurately be interpreted as safe "trophy" investments. In Chile, the PPIAF lists three different local companies as holding a total of 6 water concessions, but two of them have sold their holdings to northern private equity funds. The other Chilean interest on the PPIAF list is the Essan concession, in Antofagasta, in the northern mining area of Chile, which is actually held by Antofagasta plc, one of the oldest companies listed on the London stock exchange. This company originated as a 19th century British railway venture in Chile and Bolivia, and is now overwhelmingly a copper mining group, for whom the water concession in Antofagasta represents only 2.5% of their total sales.[9] The list also includes the Malaysian company PuncakNiaga, with contracts in Kuala Lumpur and Selangor state, which are being renationalised in the Malaysian review of the sector, and a Malaysian engineering company, Salcon, the great majority of whose activity is in traditional engineering work, with just one contract in China. Another company on the PPIAF list, Tata, is the second largest private corporation in India and one of the largest multinationals in the world, which covers many sectors including steel, engineering, IT, energy, and services – but water is not mentioned by the group even as one element of its services activities. Tata operates water supply services in only one place, Jamshedpur, which is a

[9] http://www.antofagasta.co.uk/interior/press/f_2009.html.

company town, effectively owned by Tata since 1907,[10] including the water and electricity company, JUSCO. The new owners of the Jakarta and Manila water companies, also cited by PPIAF, include a subsidiary of Jardine Matheson, the British 19th century multinational at the centre of the opium trade, and an investment fund set up in Singapore by two companies of obscure parentage from another tax haven, the British Virgin Isles, a joint venture partly owned by another fund based in the tax haven of Bermuda.[11] Manila Water, most heavily promoted as a southern company, includes as shareholders UK and Japanese multinationals and the World Bank itself, through the IFC – the main driver of the original privatisations in 1998.[12] The company has a declared policy of expanding internationally, but by 2012 its only significant overseas operation was a half share in a water treatment plant concession in Vietnam, formerly operated by Suez.[13] The only significant water company in Latin America, an Argentinian firm called Latinaguas, shows the limitations of local companies: it has been warned, criticised and or fined for underinvestment or poor customer service in 2 of its 3 concessions, and supported by public subsidies in the third; and it has expanded internationally only in a joint venture in a small town in Peru. (Marin, 2009; Hall & Lobina, 2009).

Of the above, the only southern company which can be meaningfully described as a water multinational is the Singaporean firm Sembcorp, which purchased all of Biwater's activities. As a result, it has become the owner of Biwater's former subsidiaries in a number of countries, including the UK, Chile, China, Indonesia, Panama, Philippines and South Africa.

11.4 THE FUTURE OF INTERNATIONAL WATER COMPANIES

Since the 19th century, the size and extent of the market for private operators has been determined by political decisions on public ownership and privatisation, and by the institutions driving first the empire, and later globalisation. The future prospects for this business will also depend on whether political developments locally and internationally will again move in favour of the private companies.

This seems unlikely in the near future. The 2000s saw a painful demonstration that privatised water supply is a political minefield and not as reliable a source of returns as had been expected. There is much evidence that the political resistance to privatisation is as strong as ever – for example the Italian 2011 referendum, and the continued campaigns in countries as diverse as Indonesia, Morocco and Peru. The international institutions, led by the IFC, still continue to promote water privatisation, but even those countries subject to IMF conditions, such as Greece and Portugal, may find it politically hard to privatise water operations. The international institutions themselves are also becoming less influential over development in southern countries, as China becomes more influential.

This is an unfavourable business climate for the entrance of new companies or the return of old ones. The enthusiasm of the IFIs for further privatisations is much greater than that of the companies. The English water companies have almost completely disappeared from the international scene, and it is hard to see any commercial reason for their private equity owners to re-launch them on such risky exercises. There is no capacity for serious expansion by companies such as Manila Water, even if its owners thought expansion was a risk worth taking. The decline may continue further. The Spanish multinational FCC could abandon its international presence without serious loss, in which water is a small part compared

[10]See http://www.tata.com/company/profile.aspx?sectid=zDyu5pZJHyw=; http://www.tata.com/careers/index.aspx?sectid=10thFt4D0OU=.

[11]Acuatico's takeover bid rejected by committee The Jakarta Post, Jakarta 12/16/2006; http://www.firstpacco.com/admin/upload/ir/circulars/ec061227.pdf.

[12]www.manilawater.com/files/Manila%20Water%20Bond%202013%20Prospectus%20October%209%202008.pdf.

[13]The Press Trust of India September 9, 2007.

with the core construction business. Even SAUR, the third and smallest of the French groups, is now part of a larger environmental services group, which could shed the remaining international water business without a great impact.

This leaves Suez and Veolia. The two companies are as dominant as they were within the private share of the market, but, compared with 1990, their home market has been eroded by re-municipalisations and is now smaller than it was. The private share of the world market has grown, but there are few new reliably profitable contracts outside the UK and Chile. Suez has shrunk its activity to France, Spain, the USA and a few residual contracts elsewhere, but has abandoned serious attempts to gain new water supply contracts, apart from routine expressions of hope for business in China. Veolia has been less ruthless, and even in 2011 acquired a new water supply contract in Nagpur, India, but is under shareholder pressure to shrink its operations to the same scale as Suez. Both have instead refocused on engineering work, such as water and wastewater treatment plants and desalination plants, water and sanitation services for industrial companies, and consultancy contracts. Like the companies' other main business, waste management, these are areas of reliably growing demand. It remains to be seen if private treatment plants will become as politically contentious as water supply privatisations, and if the economics of 25-year BOT contracts are more robust than those of 25-year concessions (Hall & Lobina, 2012, Suez, 2012; Veolia, 2012).

Both companies have little choice but to continue with their existing public water supply business, especially in France: the new role of the French government as owner means that both companies have to protect this dwindling domestic business from foreign ownership. Further re-municipalisation may thus become the most acceptable way for the companies to gradually divest their low-margin water business, as already has happened with energy companies in Germany.

REFERENCES

AFD (2010). Water services and the private sector in developing countries http://64.95.129.92/ppiaf/sites/ppiaf.org/files/publication/water-services-and-the-private-sector-in-developing-countries.pdf#page=30 (Consulted 02-May-2012).

Hall D. and Lobina E. (2006). Pipe Dreams PSIRU http://www.psiru.org/sites/default/files/2006-03-W-investment.pdf (Consulted 02-May-2012).

Hall D. and Lobina E. (2007). International actors and multinational water company strategies in Europe. 1990–2003. Utilities Policy, **15**(2), June 2007, 64–77.

Hall D. and Lobina E. (2009). The private sector in water in 2009. www.psiru.org/reports/2009-03-W-companies.doc. (Consulted 02-May-2012).

Hall D. and Lobina E. (2012). Water companies in Europe www.psiru.org (forthcoming) (Consulted 02-May-2012).

P. S. Juuti and T. S. Katko (eds.) (2005). *Water, Time and European cities. History Matters for the Futures*, Watertime Project, EU. http://www.watertime.net/Docs/WP3/WTEC.pdf

Lobina E. and Hall D. (2007). Water privatisation and restructuring in Latin America, 2007 http://www.psiru.org/reports/2007-09-W-Latam.doc (Consulted 02-May-2012).

Marin P. (2009). Public-Private Partnerships for Urban Water Utilities. World Bank/PPIAF Trends and Policy Options No. 8, Washington, DC, USA.

Pigeon M., McDonald D.A., Hoedeman O. and Kishimoto S. (2012). Remunicipalisation: putting water back into public hands. TNI http://www.corporateeurope.org/publications/remunicipalisation-putting-water-back-public-hands (Consulted 02-May-2012).

Suez Environnement (2012). Reference Document 2011 http://www.suez-environnement.com/wp-content/uploads/2012/04/DDR-SEC-2011-Version-anglaise-d%C3%A9finitive.pdf (Consulted 02-May-2012).

Veolia Environnement (2012). Annual report 2011 (Form 20-F) http://www.finance.veolia.com/docs/ VE-20-F-2011-EN.pdf (Consulted 02-May-2012).

Chapter 12

Issues of governance and citizenship in water services: a reflection on Latin American experiences[1]

José Esteban Castro

12.1 INTRODUCTION

Recent debates have highlighted the need for democratic, not merely "good" governance (Mokre & Riekmann, 2007). This is highly relevant for public policy in general in relation to the management of public services and particularly for water and sanitation services (WSS). In this regard, meaningful, not just formal citizen participation is at the centre of democratic governance, and it is unsurprising that "participation" has become part of the standard vocabulary in the field of public services management. There are different reasons and drivers behind this increasing interest in participation, which include from pressures exercised by social actors that seek to improve their living conditions and demand higher levels of transparency and accountability from authorities and service providers to the opportunistic manipulation of governments, businesses, and other power holders who see in "participation" an ideal mechanism for co-opting and disciplining grassroots organizations and common citizens to curb dissent and weaken opposition to unpopular and often illegitimate policy decisions. This chapter explores the concept of participation, as a component of democratic governance in the management of public services, and examines its implications for the field of WSS drawing on lessons emerging from recent experiences in Latin America.

12.2 GOVERNANCE, CITIZENSHIP AND PARTICIPATION

"Social participation" is constitutive of the central concepts of the modern Western democratic traditions such as "citizenship", "public sphere", and "civil society". However, in different cultures and contexts both the content and extension of the concept of participation, as well as its practical meaning, are wide ranging, often contradictory and even incompatible. One of the fundamental contradictions determining such variations can be traced back to the existence of rival concepts of liberty, which in turn determines

[1]An earlier version of sections of this chapter have been presented in seminars and conferences and published in Spanish by the Pan American Health Organization. Part of this work was done within the framework of the WATERLAT network (www.waterlat.org), funded by The Leverhulme Trust.

the conceptualization of different rights associated with the democratic process and particularly with citizenship rights and duties, such as the right and duty to participate in public life. For instance, while the political tradition of individualistic liberalism places the emphasis on the negative character of liberty, understood as the absence of limits or barriers to individual pursuits, in the traditions associated with radical democracy the emphasis is placed rather on the positive character of liberty. For the latter, positive liberty relates to the structural conditions that may hinder or enhance the chances of individual human beings to fully develop their potentialities, which by definition requires the existence of rules and boundaries that prevent the control or monopolization of such conditions by powerful individuals or groups of them. Between these two polar understandings of liberty there is a range of possibilities that can be identified in the field, and have a powerful influence on the development of institutional arrangements and public policy, including those involved in the management of essential public services. In particular, these diverse and often contradictory conceptions of liberty inform very different and even incompatible understandings of social participation as a citizenship right and duty.

In this connection, the concept of participation as a citizenship right is closely linked with the notion of "public sphere", which takes a diversity of forms in different territories and historical periods (Ferree *et al.* 2002). Thus, in the prevailing Western political traditions grounded on the principles of liberal representative democracy there is a tendency to reserve participation in the management of public affairs to professional politicians and experts. Contrastingly, the traditions that foster the deepening and expansion of the democratic process oppose elitist monopoly control of policy decision making and public management and demand a wider participation of the general public in the process. The deep-rooted contradictions between the elitist, restricted conception of social participation and the conceptions that understand that widening social participation is a crucial mechanism in the process of consolidation of substantive democracy, continue to strongly influence the management of public affairs, including essential public services like WSS.

The historical evidence suggests that public service management –at least in urban areas–, with few exceptions has been characteristically technocratic and hierarchical, grounded on the notion that such activities are a preserve of techno-scientific experts and professional politicians. Paraphrasing John Dryzek, in traditional management approaches the common understanding has been that these activities must be left "to the experts" (Dryzek, 1997). Somewhat paradoxically, this hierarchical model, non participatory, often paternalistic and normally closed to the scrutiny of citizens, their representatives, direct users, and the population in general has been highly successful, at least in the developed world. This is the model of public management, with different national variations, that led to the universalization of WSS and consequently to substantial improvements of living standards for large majorities over the twentieth century. There are some caveats to this argument for sure, as I am here referring to the dominant trends in Europe and the United States, but there exist many examples where the management of basic public services has taken other forms, being subject to democratic citizen scrutiny, such as in the countries of Nordic Europe characterized by strong traditions of local democracy (Pietilä *et al.* 2009). Also, it is important to recognize that the success of technocratic and non-participatory management of essential services has a relatively short history, even in the most developed countries, as the universalization of access in general took place after World War II, and in many cases since the 1960s, while in the Global South the promise of universalization was never materialized.

An uncomfortable question arises from this historical snapshot: if the dominant trends suggest that the great advances in the universalization of basic public services took place within the framework of a management model characterized by elitism, lack of participation, paternalism, and often authoritarianism, why should anyone insist in the need for participatory, more democratic management of such services? This question is connected with a much broader problem: the democratization of public services management through enhanced social participation is part and parcel of the process of democratization of society as a whole, at different levels and in different spheres of activity and responsibility. In the last analysis, the decision to support the advancement of democratization processes at the societal level, including the democratization of public services management, is a normative preference primarily grounded on the principles of equality and solidarity.

In this regard, the dominant forms of technocratic, hierarchical management, that tend to limit if not altogether exclude citizen participation in the monitoring and democratic control of how public services are run, have been criticized and confronted historically from a diversity of fronts. The experience of Latin America is exemplary in this respect, as in the last two decades the confrontation between rival conceptions of participation has been exacerbated by the far-reaching reforms introduced to promote free-market models of management in public services, often with the argument of fostering wider social participation.

12.3 TRENDS IN THE MANAGEMENT OF WSS IN LATIN AMERICA

As stated earlier, "participation" takes a diversity of forms in different contexts. It is possible to identify certain trends in the forms of participation, which are not necessarily mutually exclusive but are rather closely related and often even overlap:

1) Forms of technocratic management that exclude the participation of the population, whether in their character of citizens or as users-clients-consumers of public services;
2) Forms of technocratic management that allow (or induce) highly restricted spaces for participation, normally limited to activities such as self-help, contributions in kind, or the exercise of consumer rights;
3) Bottom-up social participation, promoted by a diversity of social organizations (workers unions, users committees, community groups, NGOs, etc.).

These are broad dominant patterns that can be identified, although in practice these trends often appear in different combinations that evolve over time as a result of the dynamics of socio-economic, cultural and policy-institutional processes. For instance, the technocratic forms of management that tend to induce restricted, controlled spaces for participation sometimes trigger unforeseen social processes that may fuel the emergence of genuinely autonomous participatory practices and institutions. Contrariwise, sometimes participatory processes that first emerge as autonomous grassroots experiences end up being co-opted and eventually demobilized or neutralized by the authorities and other power holders.

12.3.1 Technocratic, non-participatory management

As discussed earlier, technocratic, non-participatory forms of management of essential public services have been dominant. During the twentieth century, in most Latin American countries the management of WSS was fundamentally a state monopoly or, at least an activity heavily controlled by the state at different levels. In the case of Brazil, for instance, the National Sanitation Plan (PLANASA) introduced by the military dictatorship that ruled since 1964 to 1985 played a central role and continues to determine many aspects of the management of WSS in the country (Heller, 2009). Thus, although in recent years Brazil has nurtured the emergence of participatory mechanisms in the management of essential public services, the hierarchical, non-participatory model remains highly influential. As an example, in the Basin

Committees (Comités de Bacia) created in the late 1990s, which formally are composed by all water users in each basin, domestic users do not have direct participation. Rather, they are represented by the water and sanitation operators that according to the law are the "users" with the right to participate in the Committees. As a result, millions of domestic urban users are represented by the provincial water utilities created by the military dictatorship in the 1970s, which are not characterized for allowing their users participation in monitoring how WSS are managed and run. This is just an example, as in fact the situation regarding basin committees is not very different in the rest of Latin America (e.g. Castro *et al.* 2004 for the case of Mexico).

In different ways, the introduction of aggressive privatization policies in the field of essential public services since the 1990s, often justified as a form to promote greater citizen participation in monitoring the running of public services, has tended to consolidate or even aggravate the non-participatory, even authoritarian character of the technocratic model of management. The case of Argentina is an excellent example, as in the 1990s the country became an experimental field for the implementation of the policies of liberalization, deregulation, and privatization of public services, including WSS. Between 1993 and 1999 Argentina implemented a massive transfer of public WSS to the private sector mainly by granting long-term concessions (25–30 years) to multinational consortia, in most cases avoiding public debate not only with common citizens but even with their representatives in Congress.

Most concessions were granted by means of Presidential Decrees of Necessity and Urgency, as it was the case with the concession of WSS in Buenos Aires to the consortium Aguas Argentinas in 1993. Not only this concession was granted without public consultation but common citizens had no mechanisms available to monitor the management of the company, particularly its compliance with the concession contract. Even the regulatory body ETOSS (Tripartite Entity of Sanitary Works and Services) lacked independent access to the information needed to audit the company, as that information had become the private property of the concessionaire (Azpiazu *et al.* 2003). Eventually, the mounting discontent of the population owing to the failure of the private company to meet contractual arrangements led to the reform of the system and the cancellation of the contract in 2006 (Azpiazu & Castro, 2012).

Beyond the specificities of the Brazilian and Argentinean experiences discussed above, these are not isolated or extreme cases. Rather, the technocratic, hierarchical, non-participatory model of management remains very powerful in most of Latin America, even in those countries that have made significant progress in democratizing some aspects of the management of WSS in recent years (Castro, 2005). Nevertheless, in most countries of Latin America it is possible to identify a flexibilization of the conventionally rigid, strictly non-participatory or even authoritarian forms of management, whereby a limited degree of "participation", highly restricted and controlled and often induced from above, is being allowed. As a working hypothesis it can be argued that this model of management with some degree of restricted participation is now probably the most recurrent form to be identified in the region.

12.3.2 Technocratic management with restricted participation

For diverse reasons and under certain circumstances the technocratic model of management introduces elements of restricted participation, often induced and directed from above either by governments or other relevant power holders, including public or private service providers. In some cases, the process can be paternalistic, clentelistic, relatively "benign", whereby induced participation becomes a function for the political inclusion of the population, while in others it may take the form of openly manipulative mechanisms of co-option and control of social actors. Often too, the opening of restricted channels of participation is merely a concession, may be even temporary, resulting from social pressures for a greater democratization of public management. For instance, in the case of Buenos Aires commented earlier, the growing discontent of the users in the late 1990s led to the opening of additional participation mechanisms. However, these mechanisms were limited to the involvement of users as providers of materials and labour for the expansion of networks in poor areas, a programme developed by the private operator and a group of local and international NGOs (Azpiazu *et al.* 2003). Although clearly this extension of citizen "participation" represented a degree of progress with respect to the previous situation, users and citizens continued to be excluded from crucial issues such as the democratic control and monitoring of the services.

Another illustrative case comes from Mexico. In the 1981 National Water Plan, the Mexican authorities already announced the need for "wider user participation in solving common problems" (SARH, 1981). In the early 1990s the government introduced a series of reforms to create a "new water culture" grounded on citizen involvement, user responsibility, and greater private sector participation. As a result, user participation became institutionalized, for instance by creating a Coordination of Participation within the National Water Commission (CNA), a Sub-Coordination of Social Participation at the Mexican Institute of Water Technology (IMTA), and even fostering a Citizens Water Movement at the national level. However, these top-down attempts to induce and formalize user participation in water management produced neither greater or more effective participation nor the development of a "new water culture".

Despite that successive governments have insisted in the implementation of similar institutional transformations, the notion of "participation" that prevails tends to reduce the process to its technical and administrative dimensions, playing down the socio-political aspects of participation. In fact, in these government-led reforms participation tends to be understood as expected obedience from the users to decisions taken by authorities and technical experts (Torregrosa *et al.* 2003; Castro, 2006). In this model, participation does not include public debate or consultation about the principles that should guide the management of essential public services or about the type of management (public, private, community, etc.) that should be chosen, among other issues that are not open to democratic debate or scrutiny. Even more, the government decides who can participate in the newly created spaces for participation, which severely restricts the possibility of genuinely autonomous participation of users and citizens (Jiménez & Torregrosa, 2009). Too often, the participation promoted by this model of management is limited to increasing the responsibility of domestic users, particularly in accepting tariff increases (willingness to pay) but also in relation to the requirement of direct investments from the users to get connected to the services, whether through monetary contributions or through the provision of materials and voluntary labour.

A third case to exemplify the workings of the technocratic model comes from research carried out in Bolivia (Crespo Flores *et al.* 2004). The country passed a new Law of Popular Participation in 1994 to promote citizen involvement in local government affairs and the simultaneous creation of a new regulatory framework for public services included some participatory mechanisms such as public consultations. However, "participation" in these reforms was restricted to the provision of administrative channels for user complaints and appeals: when the WSS of Cochabamba were privatized in 1999 the decision was not subject to citizen scrutiny. The system was marred by lack of legitimacy and very low public trust: a public consultation called in December 1999 to discuss a tariff increase to be applied by the private operator attracted only 14 people, as a majority of the population perceived the regulator as a representative of the interests of the private company. It is well known that the privatization of WSS in Cochabamba had a violent end with the so-called "Water War", a popular mobilization that led to the cancellation of the contract and the partial collapse of the national government in March 2000.

The tense interrelation between the two models of technocratic management, non-participatory and with limited participation, was clearly manifest in the political confrontations played out in Brazil since the year 2003 around the project for a National Law of Environmental Sanitation proposed by President Luiz Inácio Lula da Silva (2003–2010). The law project was conceived in the context of a widespread popular mobilization seeking to democratize the management of public services. The mobilization was oriented by the principles of direct democracy, demanding a greater role for social organizations and citizen movements in decision making and monitoring of policy and implementation of essential public services. The initiative was led by grassroots movements such as the National Front of Environmental Sanitation (FNSA) and organizations like the National Association of Municipal WSS (ASSEMAE) some of whose members came to occupy positions in the national government, particularly in the National Secretariat of Environmental Sanitation of the Ministry of the Cities. The law project faced powerful resistance from the political opposition and business and corporate lobbies with strong interests in WSS. Eventually, the mechanisms of direct democracy originally envisaged by the grassroots movements that contributed to elaborate the project were severely restricted or even excluded in the final National Law of Basic Sanitation that was passed in 2007. This outcome illustrates the enduring prevalence of technocratic, hierarchical and non-participatory management approaches. It is also an example of the political dimension of management activities, which are often understood as merely technical-administrative affairs but that in practice constitute an arena of confrontation between rival, often incompatible understandings of democracy, citizenship, and participation.

12.3.3 Bottom-up social participation

In recent decades there has been a widespread social mobilization oriented at advancing the process of democratization of the management and access to essential public services, notably WSS and the collection and recycling of solid waste in most Latin American countries (Grosse *et al.* 2004; Cárdenas *et al.* 2005; Medina, 2005; Grosse *et al.* 2006; Bell *et al.* 2009; Castro, 2008; CEDA, 2009; Red Latinoamericana de Recicladores, 2010). This mobilization has taken different forms, from denunciations and pacific demonstrations to violent confrontations, many times resulting in the loss of human life. Very often, in the face of the inaction of the state, grassroots actors participating in these mobilizations have decided to take responsibility for the organization of essential services in their communities, for instance through self-organized community systems or local or regional cooperatives. In some cases grassroots movements have showed great capacity for action and political articulation, which has allowed them to mobilize significant political and other resources and strengthen their capabilities for direct intervention in the management of these services, as it has happened among other examples in Argentina, Bolivia, Brasil, Ecuador, Mexico, Nicaragua, Uruguay and Venezuela. The cases discussed below represent highly successful forms of bottom-up social participation, though some were often later co-opted or demobilized, sometimes as a result of their success, for instance because the leading social actors involved were part of wider political movements that eventually came to control regional or national governments.

The first case is again related to the Bolivian experience already discussed. The Water Wars, first in Cochabamba (1999–2000) and later in La Paz-El Alto (2005–2006), became a global emblem of popular struggles against the privatization of essential public services. In particular the case of Cochabamba had wide repercussions both at the national level when the entire government resigned in March 2000, with exception of the President Hugo Banzer Suárez (1997–2002), and at the international level, where it came to symbolize the movements of popular participation for the democratization of the management of essential public services (see among others: Crespo Flores *et al.* 2004; Laurie & Crespo, 2007; Spronk & Webber, 2007; Pérez Barriga, 2010). In Bolivia, many of the organizations that led the mobilization against privatization were a fundamental component of the social base that eventually brought President Juan Evo Morales Ayma to power in 2006 and would later have a significant role in the attempt to reorganize the management of public services in the country.

The second case covers several less well-known examples from Argentina. One example concerns citizen struggles against the privatization of WSS in the province of Tucumán during the 1990s. In 1995 the provincial government granted a long-term concession to run WSS to a multinational private operator. The project was affected from the start by allegations of corruption, lack of transparency, and absence of public consultation and debate, which triggered strong resistance. This was aggravated when the first decision of the privatized company was to raise the tariff by 105% (including a tax to fund the regulator). These events prompted massive citizen protests that included among the participants municipal authorities, provincial legislators, workers that had been laid off by the privatized company, and representatives of local businesses. Around 86 percent of users participated in a campaign of civil disobedience consisting in the nonpayment of the water bills, public demonstrations, and a "Popular Congress" to denounce the alleged corruption of the privatization process. Eventually the privatization contract was canceled in 1997 (Crenzel, 2004).

Another example is from the province of Chaco. In 1994, in compliance with the provincial Constitution, the government called a Public Consultation about plans to privatize public services promoted by the national authorities. The most important political parties in the province were in favour of privatization but the political establishment was shocked when a majority of voters rejected the privatization plans. The Public Consultation was legally binding, but regrettably the sovereign decision of Chaco's citizens was severely penalized by the national government, which excluded the province from the investment programme in public services infrastructure. The funding for this programme was provided by the Inter American Development Bank (IDB) and the privatization of public companies was a condition to receive the funding (Roze, 2002).

Other important examples of grassroots participation took place in Buenos Aires, where the population switched from apathy in the early 1990s to the very active participation that put pressure on the authorities and the privatized companies and eventually led to the cancellation of the concession contracts for WSS (Azpiazu *et al.* 2003). Similarly, in the provinces of Santa Fe and Córdoba environmental groups, workers unions, and community organizations, with the support of academics and local politicians implemented very effective campaigns demanding the cancellation of the concession contracts to provide WSS in both provinces. In Santa Fe the popular mobilization, helped by the ascension to power of a center-left alliance, succeeded and in 2006 the privatization contract was cancelled, the services were placed back in public hands, and the provincial government implemented a number of policies to foster citizen participation in monitoring the management and performance of the water utility (Rovere, 2010). In Córdoba at the time of writing this chapter grassroots movements continue a long-term struggle demanding the cancellation of the privatization contract, the devolution of WSS to the public sector, and the search for practical solutions to existing problems of access and quality through self-organization in the poorer areas of the capital city (Spedale, 2009; Berger, 2010). In addition to the cases of WSS briefly described here, there exist significant examples of successful forms of bottom-up social participation in the self-organization of other basic services, notably the collection and recycling of solid waste (see, for instance: Paiva, 2004; Angélico & Maldovan, 2008; Los Caminantes, 2010).

The third case is from Brazil, which is perhaps one of the internationally best known and celebrated examples of bottom-up social participation, connected with the development of innovative mechanisms of direct democracy. Among other bottom-up initiatives introduced is worth noting the well-known experience of Participatory Budgeting, initiated in the city of Porto Alegre, an experience that came to constitute a paradigmatic example of participatory policy (Dutra & Benevides, 2001). Figure 12.1 shows one of these meetings where around 2000 people from a poor neighbourhood in the city of Recife, Pernambuco, gathered to discuss local needs with the municipal authorities and set the investment priorities for their local area.

Figure 12.1 Neighbourhood meeting of the municipal Participatory Budgeting, Recife, Brazil, July 2009. Author's collection.

Another example are the participatory activities promoted by the Councils and Conferences of the Cities organized by the government to foster public debate and setting policy priorities for local services. International institutions like the World Bank have recognized the value of such initiatives to mobilize the population and generate legitimacy in decision making processes (World Bank, 2003:42). In the particular case of WSS Brazil has numerous experiences of successful public management, often at the local level, with a substantial degree of social participation (Miranda Neto, 2005; Costa *et al.* 2006). Similar examples of highly successful bottom-up participatory and politically inclusive processes can be found in relation to the movements of "garbage collectors" (catadores de lixo) and "environmental agents" (Pimenta Velloso, 2005; MNCR, 2010).

I will refer here to two additional examples that illustrate the emergence of bottom-up, autonomous participatory forms of management of essential public services in Latin America. One case is from Nicaragua (Barrios Jackman & Wheelock Díaz, 2005; Kreimann Zambrana, 2009), where since the 1990s government policies sought to decentralize WSS by transferring them to local governments. In some places these policies were actually a response to popular mobilizations demanding greater local control over essential services. The government also promoted formal mechanisms of social participation such as Open Councils, Consumer Associations and Committees of Municipal Development. However, recent research suggests that the most effective forms of social participation are related mainly to two different bottom-up experiences: (a) the management of essential services in areas not served by the state, especially in rural and peri-urban areas, and (b) the struggle against the privatization of public companies.

These participatory experiences are connected with a long-standing tradition of popular organization in the country that can be traced back to the 1960s with the creation of "water committees" to solve the problem of access in peripheral urban settlements. Today Nicaragua has around 6000 Potable Water and Sanitation Committees (CAPS) working in rural and peripheral urban areas, serving around one quarter of the country's population. Many of these CAPS have reportedly developed effective mechanisms for social participation and democratic control, are composed by members elected by the local communities and have responsibility for the collection of water fees and the maintenance of services. CAPS members are elected in Community Assemblies, which also serve to consult the population on issues such as tariff increases, investment priorities, regulation of water uses, and so on. However, the relative autonomy of operation enjoyed by the CAPS in relation to the government and other actors is also the object of significant tensions as local self-organization often clashes with processes commanded from the central government. Like in many other countries of the region bottom-up participatory processes were boosted by widespread popular rejection of privatization policies, in this case particularly since 2001. A crucial moment was the creation of the Alliance Against Privatization and for the Access to Water in 2003, which brought together existing movements organized by the Churches Network for Joint Action and NGOs representing consumer rights. These movements are highly heterogeneous, but they have found a common ground in the defense of water as common good and essential WSS as a public good and a human right.

The second and final example is from Venezuela (see: Arconada Rodríguez, 2005, 2006; Lacabana & Cariola, 2005; López Maya, 2008). Like in other Latin American countries participatory processes around the management of basic services in Venezuela can be traced back at least to the 1960s. In recent times, one of the foundational experiences took place in the 1990s in the peripheral areas of the capital Caracas, where local parishes started the organization of bottom-up processes of direct democracy connected with the management of WSS. These processes led to the creation of Technical Water Boards (MTAs) and Parish Governments, self-organized bodies to facilitate popular participation, in the municipalities of Antímano and El Valle. These developments would later become a blueprint for national policies implemented after the arrival of President Hugo Chávez Frías in 1999. The experiences of these two municipalities was used as a trigger for a public debate that eventually led to the elevation of MTAs to national policy within the framework of the new country's constitution.

Also inspired by the bottom-up experiences, the national government created Community Water Councils, and later Community Self-Managed Organizations (since 2004) to promote effective social participation in issues of local interest such as the state of river basins and essential public services. Among other concrete participatory mechanisms, the MTAs bring together local members of the community, local politicians and technical experts from the public water utilities to jointly map the local situation, which includes the elaboration of graphical representations of the infrastructure networks called "sketches" (croquis). The participatory sketches reflect situations entirely unknown to the authorities and service providers, including elements of the WSS infrastructure such as local networks, master valves and other elements that had been lost from the official records. In addition, the MTAs develop diagnoses and intervention projects, monitor the implementation of the infrastructure works and service quality, co-manage funding jointly with local authorities, and help with the collection of service fees. Also, the MTAs represent their regions in regular meetings of the Community Water Councils, which serve to coordinate the activities of the MTAs in whole districts. It is estimated that there are currently around 2700 MTAs operating in the country.

12.4 OBSTACLES AND POSSIBILITIES FOR THE SUBSTANTIVE DEMOCRATIZATION OF WSS MANAGEMENT

The above examples not only illustrate some of the most important trends in participatory management of WSS in Latin America, but also contribute to identify some of the obstacles and opportunities facing the democratization of management systems. The social struggles oriented at widening democratic spaces that can be identified in Latin America have opened opportunities for transformation and allowed important achievements in several dimensions, from the introduction of direct democracy in the management of WSS in large cities to the consolidation of autonomous management systems in rural and periurban areas. These experiences suggest that there is great potential for the expansion of social participation as a mechanism to democratize management systems that have been historically organized as hierarchical, elitist, and even authoritarian structures. However, there are no panaceas and bottom-up participatory processes, even when they genuinely emerge from the grassroots, are exposed to all manners of distortions and threats, from corruption and internal degradation of the democratization process when the main actors come to occupy political positions to co-optation and manipulation of social movements by governments, international cooperation and financial agencies, or multinational corporations (Cooke & Kothari, 2001). Often too, governments have promoted forms of social participation to cover up their reluctance to take responsibility for ensuring the universal access to essential public services. They do this sometimes by transferring the financial burden of providing basic infrastructure to the population, normally to unserved sectors that also tend to be the poorest and most marginalized.

From a general perspective, the prevalence of forms of representative democracy based on the principle that the management of public affairs is a matter for professional politicians and technical experts constitutes a central obstacle for the advancement of democratization processes. These forms of representative democracy continue to occupy a prominent place in Latin America, even in the countries that have experienced significant transformations fuelled by the widespread mobilization of grassroots actors such as Argentina, Bolivia, Brazil, Ecuador, Uruguay or Venezuela. This situation was aggravated by regressive tendencies experienced since the late 1980s, which further exacerbated the conditions of exclusion affecting a large proportion of the population. However, in Latin America there are deep-rooted traditions of social struggle whereby grassroots movements have historically confronted the state and other power holders in the search for more dignified living conditions and the advancement of substantive, not merely formal or rhetorical democracy. The persistence of these traditions and the emergence of new forms of bottom-up social participation in recent years help to explain some of the progressive transformations experienced in the region since the beginning of the Twentieth-first Century. This is particularly true about the opening of channels for social participation in public affairs, although these tend to be limited, circumscribed to specific spheres of activity and well-defined territories, time bound, and normally closely monitored or controlled by the state. In the case of essential public services, governments frequently respond to social pressures by promoting forms of social participation restricted to activities of self-help and co-responsibility involving the provision of financial resources, materials, or labour in order to extend infrastructure networks into unserved areas. However, with few exceptions, more substantive questions such as decisions about whose material interests should receive priority or what values and principles should inform the management of WSS are consistently excluded from public debate and are closed to the participation of common citizens.

This kind of social participation, limited, restricted, closely surveilled, bound to specific issues, spaces and timespans and that leaves out the issues that are politically substantive is also promoted by international cooperation agencies and financial institutions, which consider "participation" as a key prerequisite for the granting of investment funding for the infrastructure of essential public services. Regrettably, too often the understanding of "participation" that prevails among these actors could be translated as "expected anticipated obedience" to the decisions taken by professional politicians and technical experts on behalf of the beneficiary population. Substantive citizen participation in decision making and democratic control and scrutiny of the management of public services is out of the question, with rare exceptions. In recent years the influence of this prevailing, top-down, understanding of participation has become clearly manifest, for instance in relation to the commodification of essential public services, and there is a clear trend towards the abandonment of the notion that essential public services constitute a public good or a citizenship right, not to say a human right. Although the most aggressive initiatives informed by this instrumental notion of participation in the field of WSS have been related to privatization policies, the process of commodification of essential services is taking place regardless of the public or private character of the service providers. Public companies are increasingly pressured to adopt market criteria for their operation, and therefore the social objectives that should inform the provision of essential services are increasingly subordinated to the logic of the market: making "profit" is increasingly higher than providing a safe and universal service in the hierarchy of priorities set for public operators, which are required to perform as private enterprises. There has been an erosion of the ethics of the public space, which despite all its shortcomings had informed the development of essential public services since the late nineteenth century that led to the universalization of access in the most advanced countries.

This erosion of the ethics of the public good, that was at the base of the notion that access to goods and services essential for a dignified life must be independent from the capacity to pay of individuals and families, and its replacement for a mercantilist ethics that prioritizes profit over rights is one of the major obstacles facing the substantive democratization of the management of essential public services. Defending and claiming back the ethics of the public good and expanding and deepening its meaning to incorporate the principle that access to services essential for human life constitute a common good whose management must be excluded from the logic of the market constitutes a fundamental challenge facing human civilization in the Twentieth-first Century.

REFERENCES

Angélico H. and Maldovan J. (2008). *El reciclaje de residuos sólidos urbanos: las cooperativas como un actor diferenciado en el circuito productivo* (The recycling of urban solid waste: the cooperatives as a specific actor in the productive circuit). V Meeting of Latin American Researchers on Cooperativism, Ribeirão Preto, São Paulo, Brazil: Alianza Cooperativa Internacional.

Arconada Rodríguez S. (2005). Seis años después: mesas técnicas y consejos comunitarios de agua [...] (Six years later: water technical boards and community councils). *Revista Venezolana de Economía y Ciencias Sociales*, **11**(3), 187–203.

Arconada Rodríguez S. (2006). Mesas técnicas de agua y consejos comunitarios de agua (Water technical boards and community councils). *Revista Venezolana de Economía y Ciencias Sociales*, **12**(2), 127–132.

Azpiazu D. and Castro J. E. (2012). Buenos Aires in muddled waters. In: Remunicipalisation. Putting Water back into Public Hands, M. Pigeon, D. A. McDonald, O. Hoedeman and S. Kishimoto (eds), Aguas públicas, Amsterdam, Transnational Institute, pp. 58–73.

Azpiazu D., Catenazzi A., Crenzel E. A., da Representaçao N., Forte G., Forcinito K. and Marínet J. C. (2003). Buenos Aires – Argentina case study report. In: PRINWASS, Research Project, J. E. Castro (Coord.), European Commission, Fifth Framework Programme, INCO-DEV, Contract PL ICA4-2001-10041, University of Oxford, Oxford.

Barrios Jackman M. and Wheelock Díaz S. I. (2005). Movilización Social y Gestión del Agua en Nicaragua (Social Mobilization and Water Management in Nicaragua). Universidad Centroamericana (UCA), Managua.

Bell B., Conant J., Olivera M., Pinkstaff C. and Terhorst P. (2009). Changing the Flow: Water Movements in Latin America. Food and Water Watch, Other Worlds, Reclaiming Public Water, Red VIDA, Transnational Institute, Washington, DC.

Berger M. (2010). Cuerpo, experiencia y narración: las prácticas de autoorganización en la defensa de derechos. [...] (Body, Experience and Narrative: Self-Organization in the Defense of Rights). PhD in Social Sciences, Faculty of Social Sciences, University of Buenos Aires, Argentina.

Cárdenas N., Makovski P., Rodríguez E., Vargas L., Vargas R. and Cordero M. (2005). *Construyendo Ciudadanía: el Derecho Humano al Agua* (Constructing citizenship: the human right to water). *Consorcio de Investigación Económica y Social (CIES), Observatorio del Derecho a la Salud, Centro de Estudios para el Desarrollo y la Participación (CEDEP)* (Consortium of Economic and Social Research (CIES), Observatory of the Right to Health, Center for the Study of Development and Participation (CEDEP)), Lima.

Castro J. E. (2005). Agua y gobernabilidad: entre la ideología neoliberal y la memoria histórica (Water and governance: between neoliberal ideology and historic memory). *Cuadernos del CENDES*, **22**(59), 1–22.

Castro J. E. (2006). Water, Power, and Citizenship. Social Struggle in the Basin of Mexico. Palgrave-Macmillan, Houndmills, Basingstoke and New York.

Castro J. E. (2008). Water struggles, citizenship and governance in Latin America. *Development*, **51**(1), 72–76.

Castro J. E., Kloster K. and Torregrosa M. L. (2004). Ciudadanía y gobernabilidad en México: el caso de la conflictividad y la participación social en torno a la gestión del agua (Citizenship and governance in Mexico: conflict and social participation around water management). In: El Agua en México Vista desde la Academia (Water in Mexico: an Academic Perspective), B. Jiménez and L. Marín (eds), Academia Mexicana de Ciencias, Mexico City, pp. 339–370.

Centro Ecuatoriano de Derecho Ambiental (CEDA) (2009). Memorias del Foro "Los nuevos retos de la Participación Ciudadana en el Ecuador: una mirada desde lo socioambiental (Memories of the Forum "New Challenges for Citizzen Participation in Ecuador […]). CEDA, Quito.

Cooke B. and Kothari U. (eds). (2001). Participation: the New Tyranny? Zed Books, London.

Costa S. S. Da, Heller L., Moraes L. R. S., Borja P. C., de Melo C. H. and Sacco D. (2006). Successful Experiences in Municipal Public Sanitation Services from Brazil. Transnational Institute and National Association of Municipal Services of Water and Sanitation (ASSEMAE), Amsterdam.

Crenzel E. A. (2004) Tucumán, Argentina, case study. In: PRINWASS, Research Project J. E. Castro (Coord.), European Commission, Fifth Framework Programme, INCO-DEV, Contract PL ICA4-2001-10041, University of Oxford, Oxford.

Crespo Flores C., Laurie N. and Ledo C. (2004). Cochabamba, Bolivia, case study. In: PRINWASS, Research Project, J. E. Castro (Coord.), European Commission, Fifth Framework Programme, INCO-DEV, Contract PL ICA4-2001-10041, University of Oxford, Oxford.

Dryzek J. S. (1997) The Politics of the Earth. Environmental Discourses. Oxford University Press, Oxford.

Dutra O. and Benevides M. V. (2001). Orçamento Participativo e Socialismo (Participatory Budgeting and Socialism). Perseu Abramo, Sao Paulo.

Ferree M. M., Gamson W. A., Gerhards J. and Rucht D. (2002). Four models of the public sphere in modern democracies. *Theory and Society*, **31**(3), 289–324.

Grosse R., Santos C., Taks J. and Thimmel S. (eds) (2006). Las Canillas Abiertas de América Latina II. La Lucha contra la Privatización del Agua y los Desafíos de una Gestión Participativa y Sustentable de los Recursos Hídricos (The Open Taps of Latin America II. The Struggle agains Water Privatization and the Challenges of Partipatory and Sustainable Management of Water Resources). Casa Bertolt Brecht, Montevideo.

Grosse R., Thimmel S. and Taks J. (eds) (2004). Las Canillas Abiertas de América Latina II. La Lucha contra la Privatización del Agua y los Desafíos de una Gestión Participativa y Sustentable de los Recursos Hídricos (The Open Taps of Latin America II. The Struggle agains Water Privatization and the Challenges of Partipatory and Sustainable Management of Water Resources). Casa Bertolt Brecht, Montevideo.

Heller L. (2009). Water and sanitation policies in Brazil: historical inequalities and institutional change. In: Water and Sanitation Services. Public Policy and Management, J. E. Castro and L. Heller (eds), Earthscan, London and Sterling, VA, pp. 321–337.

Jiménez B. and Torregrosa M. L. (2009) Challenges facing the universal access of water and sanitation. In: Water and Sanitation Services. Public Policy and Management, J. E. Castro and L. Heller (eds), Earthscan, London and Sterling, VA, pp. 338–352.

Kreimann Zambrana M. R. (2009). *Gestión social de un bien común: Los Comités de Agua en Nicaragua* (Social management of a common good: the Water Committees in Nicaragua). MsC in Urban Studies, El Colegio de México, Mexico City.

Lacabana M. and Cariola C. (2005). *Construyendo la participación popular y una nueva cultura del agua en Venezuela* (Constructing popular participation and a new water culture in Venezuela). *Cuadernos del CENDES*, **22**(59), 111–133.

Laurie N. and Crespo C. (2007). Deconstructing the best case scenario: lessons from water politics in La Paz-El Alto, Bolivia. *Geoforum*, **38**(5), 841–854.

López Maya M. (2008). Innovaciones participativas en la Caracas bolivariana: La MTA de la pedrera y la OCA de barrio Unión-Carpintero (Participatory innovations in Bolivaria Caracas [...]). *Revista Venezolana de Economía y Ciencias Sociales*, **14**(1), 65–93.

Los Caminantes (2010). Cooperativa de recolectores de basura (Cooperative of garbage collectors), Retrieved 9 May 2010, 2010, from http://www.loscaminantescoop.com.ar.

Medina M. (2005). Cooperativas de recicladores informales en América Latina (Cooperatives of Informal Garbage Recyclers in Latin America). Asociación Interamericana de Ingenieria Sanitaria y Ambiental (AIDIS), División Técnica de Residuos Sólidos (DIRSA), Mexico City.

Miranda Neto A. (2005). Recife, Brazil: building up water and sanitation services through citizenship. In: Reclaiming Public Water. Achievements, Struggles and Visions from Around the World, B. Balanyá, B. Brennan, O. Hoedeman, S. Kishimoto and P. Terhorst (eds), Transnational Institute and Corporate Europe Observatory, London, pp. 113–119.

Mokre M. and Riekmann S. P. (2007). From Good Governance to Democratic Governance? A Policy Review of the First Wave of European Governance Research. European Commission, Luxembourg.

Movimento Nacional dos Catadores de Materiais Recicláveis (MNCR) (2010). (National Movement of Collectors of Recyclable Materials), Retrieved 9-10-2010 from http://www.mncr.org.br.

Paiva V. (2004). *Las cooperativas de recuperadores y la gestión de residuos sólidos urbanos en el área metropolitana de Buenos Aires* (The cooperatives of waste collectors and the management of urban solid waste in the Buenos Aires Metropolitan Area). *Revista Theomai*, (Special issue) http://redalyc.uaemex.mx/pdf/124/12499309.pdf (accessed 16 August 2012).

Pérez Barriga J. (2010). Challenges and Opportunities for the Participatory Management of Water and Sanitation Services. The Case of La Paz-El Alto, Bolivia. MPhil in Geography, School of Geography, Politics and Sociology, Newcastle University, United Kingdom.

Pietilä P. E., Gunnarsdóttir M. J., Hjorth P. and Balslev Nielsen S. (2009). Decentralized services: the Nordic experience. In: Water and Sanitation Services. Public Policy and Managament, J. E. Castro and L. Heller (eds), Earthscan, London and Sterling, VA, pp. 218–233.

Pimenta Velloso M. (2005). *Os catadores de lixo e o proceso de emancipação social* (The garbage collectors and the process of social emancipation). *Ciência & Saúde Coletiva* (Science and Collective Health), **10**(supp), 49–61.

Red Latinoamericana de Recicladores (2010). (Latin American Network of Recyclers), Retrieved 9-5-2010 from http://www.redrecicladores.net.

Rovere L. (2010) Historia de la Asamblea Provincial por el Derecho al Agua (History of the Provincial Assembly for the Right to Water). Taller Ecologista, Rosario, Argentina.

Roze J. P. (2002). Chaco, Argentina, case study. In: PRINWASS, Research Project, J. E. Castro (Coord.), (European Commission, Fifth Framework Programme, INCO-DEV, Contract PL ICA4-2001-10041), University of Oxford, Oxford.

Secretariat of Agriculture and Water Resources (SARH) (1981). Plan Nacional Hidráulico (National Water Plan). SARH, Mexico City.

Spedale G. (2009). Córdoba: crisis hídrica y shock social (Cordoba: Wáter Crisis and Social Shock). Coordinadora Córdoba en Defensa del Agua y la Vida (CCODAV), Córdoba, Argentina.

Spronk S. and Webber J. R. (2007). Struggles against accumulation by dispossession in Bolivia: the political economy of natural resource contention. *Latin American Perspectives*, **34**(2), 31–47.

Torregrosa M. L., Saavedra F., Padilla E., Quiñones A., Kloster K., Cosío G. and Lenin C. (2003). Aguascalientes, Mexico, case study. In: PRINWASS, Research Project, J. E. Castro (Coord.), (European Commission, Fifth Framework Programme, INCO-DEV, Contract PL ICA4-2001-10041), University of Oxford, Oxford.

World Bank (2003). World Development Report 2004: Making Services Work for Poor People. World Bank, Washington, DC.

Section IV

Long-term policies

Section IV, Long-term policies has four chapters.

The 1st chapter was written by two authors, Blokland M. & Schwartz K. Their focus is the upscaling and internationalisation of the Dutch water supply sector.

The 2nd chapter was written by one author, Nyangeri E. N. He examines water services management and governance in Kenya, and tries to find past lessons that could contribute to a sustainable future.

The 3rd chapter was written by one author, Nilsson D. He writes about the curse of novelty and asks: How can water sector donors learn from experience?

The fourth chapter was written by two authors, Juuti P. & Rajala R. They explain the developments related to the reforming of water services in the Helsinki Metropolitan Area, Finland, a few years ago.

Chapter 13

Upscaling and internationalization of the Dutch water supply sector

Maarten Blokland and Klaas Schwartz

13.1 INTRODUCTION

Initial development of drinking water supply in the Netherlands was left to local initiative. Triggered by repeated cholera epidemics that ravaged the Netherlands from the 1830s onward, the 1867 "Report to the King" describes the generally poor condition of water supply throughout the country and emphasises the need for a national initiative. Despite this advice it would be another 40 years before the national government became involved. Until then local governments and private entrepreneurs established piped water supply systems. They did so particularly in the larger and richer municipalities, where attractive rates of return on investment could be achieved. The necessary capital was made available either from municipal budgets or provided by local, British and Belgian private financiers. By comparison, the provision of the rural municipalities lagged behind. As a result, the initial development of water supply in the Netherlands was marked by a strong urban bias (Figure 13.1).

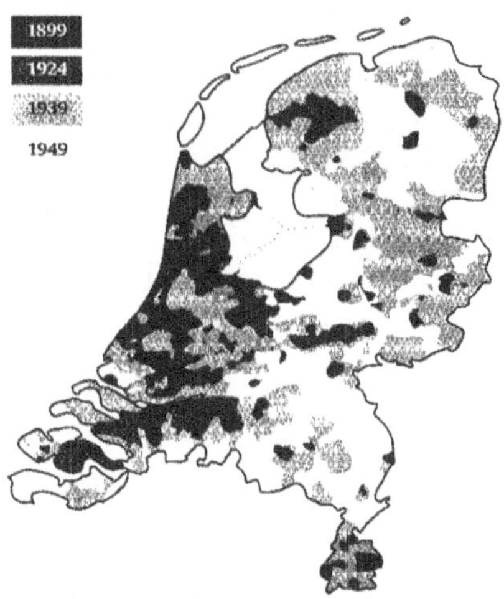

Figure 13.1 Urban-biased development of coverage in the Netherlands.

From 1910 onwards the situation started to change. Funds were allocated to water supply at the national level, and in 1913, a permanent advisory committee to the government and a national bureau were established to advise on and assist with drinking water supply development. Given the urban bias of water supply development, their concern was mainly with the development of rural, particularly regional, water supply systems. These national initiatives created the necessary administrative and professional capacity at the supra-municipal level to further the extension of water supply coverage to the rural areas – a development that would take more than 50 years to complete.

Since the initial decades of water supply provision, the Dutch water supply sector has undergone marked changes. Since the late 1930s the scale of operations and the size of the water utilities have been increasing. In 1938 the Netherlands had 232 water utilities serving less than 9 million inhabitants while currently no more than 10 water utilities supply close to 17 million people with water. Underlying this seemingly stable trend of scale increase, however, are a myriad of changes in the water supply sector. These changes pertain to the predominant objectives guiding development of the water supply sector, organizational forms of service provision, the institutions under which the water supply companies operate, and the increasingly international focus of the utilities. In this chapter we explain how these changes guided the sector to increasingly centralized service provision over the past 80 years.

13.2 AGGREGATION TO REACH UNIVERSAL SERVICE EXPANSION

Following the Second World War, the Netherlands was characterized by rapid economic development and population growth. These developments were reflected in water use which almost quadrupled between 1945 and 1970. At the same time, the country still had large rural areas unserved by the water utilities. Achieving universal service coverage in the Netherlands was the guiding objective of the water services sector for a large part of the 20th century. The drive to realize this objective was essentially two-fold. First of all, the national and provincial governments implemented licensing policies which prioritized license-applications that included service expansion to rural areas. At the same time, the national government provided subsidies for expanding infrastructure to remote villages. The national subsidy covered the non-viable part of investment capital, leaving the viable part and all operational costs to be recovered in full from the customers of the water utility. Secondly, national legislation steered utilities to mergers. In 1957, the first Drinking Water Supply Act came into force. Apart from laying down sectoral planning guidelines and water quality regulations, the Act also required the reorganisation of the drinking water sector into larger units that would be able to exercise quality control and face new technical and commercial challenges. Larger regional companies were preferred over municipal utilities because the former were able to spread out the costs of providing services to remote areas over more municipalities and customers (Blokland, 1999).

In this era of service expansion to rural areas, two additional factors aided the upscaling of service provision. Firstly, the development of the road network and telecommunications infrastructure greatly enlarged the area that utility offices could cover. For example, in 1932 the water utility in the province of Friesland covered 16 municipalities through eight district offices. By the end of the 1980s, when the service area had been expanded to cover all 44 municipalities, the number of district offices had been reduced to five. While much of materials for expanding the service network in the 1930s were transported over water, in the 1950s, with the improvements in the road network, road transportation became much more important. The improved road network and telecommunications infrastructure also allowed operational staff to cover a much larger area. The result was that the operational activities were upscaled and became more centralized. Secondly, the organizational setup of the regional utilities supported their upscaling. After experimenting with a number of organizational forms including associations and cooperatives, the majority of utilities chose to become government-owned private companies. The essential feature of a government-owned private company is that the company is established and operates under private company law whilst the shares of the company are exclusively in the hands of the national, regional or local government. Generally, there is no legal or organizational difference between a publicly- and a privately-owned company apart from the government ownership of shares. This organizational structure made it relatively easy to expand the utility to cover additional municipalities which could easily become shareholders of the company without changes in the organizational structure.

13.3 FROM UNIVERSAL COVERAGE TO EFFICIENCY GAINS

Having achieved universal service coverage by the 1970s, the main objectives of the water sector started to change. Rather than emphasizing service expansion, efficiency gains became the predominant objective. The strategy for achieving efficiency gains was similar to that for expanding service coverage: a scale increase. The Dutch national government maintained pressure on the sector to increase the scale of the utilities as it had concluded that for a water supply company to be able to meet its future supply challenges, a certain scale had to be achieved. The national government decided that a water supply utility should have (Dane & Warner, 1999):

- Appropriate management, expertise and organization
- A laboratory for quality control
- At least 100,000 connections or a supply of five million m^3 of water per year in order to produce potable water on a larger and more efficient scale.

Despite this policy, municipalities seemed reluctant to have their municipal utilities merge with other municipal utilities. As the pace of upscaling lagged behind expectations, the Dutch government eventually decided to empower the Provinces to reorganize their water supply sector in case sufficient progress was not achieved. In 1975, the section of the law on reorganisation was amended to include this provision. The amendment resulted in a wave of mergers, which swept through the water supply sector in the 1970s and 1980s reducing the number of companies from 185 in 1965 to 10 by 2008. The operational characteristics of these companies reflected the scale increase. While in 1994 the average company had 145,000 connections, in 2008 it had more than 750,000 connections.

While the number of connections grew continuously from 5.7 million in 1990 to more than 7.5 million in 2008 (+33%), the number of employees declined from 8400 to 4900 over the same period (−41%).

Another factor besides the push for efficiency from the national government intensified the efficiency-focus of the Dutch drinking water sector. Following the privatization of the English and Welsh drinking water sectors in 1989, and the start of the Buenos Aires Concession in 1993, the topic of privatization and independent regulation featured prominently in the global water community. During this era many countries either proposed or implemented reforms with the aim of increasing private sector involvement. The Dutch water supply sector was no exception. The debate mainly involved the Ministry of Economic Affairs, which favoured increasing private sector involvement and the establishment of an independent regulator, and the Ministry of Housing, Spatial Planning and the Environment, which preferred the existing organizational set-up and system of self-regulation[1]. The apex of the debate occurred in 1997–1998, when a report of the Ministry of Economic Affairs concluded that privatization could reduce water tariffs by 10% (Dijkgraaf & van de Mortel, 1997). By 1999, however, the Minister of Housing,

[1]The different perspectives of the two Ministers reflect the political parties to which they belonged. The Ministry of Economic Affairs was headed by a Minister from the Liberal Party whist the Ministry of Housing, Spatial Planning and the Environment was headed by a politician from the Labour Party.

Spatial Planning and Environment persuaded the Dutch Cabinet to prohibit privatization of the Dutch public drinking water supply sector. Liberalization was limited to the supply of water for industrial usage. Despite the fact that the water utilities were to remain in public hands and self-regulating, the debate between the two Ministries did put pressure on the water utilities to emphasize efficiency in their operations. Under such pressure, most utilities decided that further scale increases were desirable. The mergers were argued on two main points:

(1) The water utilities were too small to react adequately to new technological developments in the provision of water services. Developing new water treatment technologies required considerable investments in innovation and the Dutch water utilities were considered too small to cover these costs by themselves.

(2) Upscaling the activities of utilities which are prone to economies of scale would lead to increasing efficiency (and thus to cost savings). Especially increased automation, computerization, outsourcing and customer management were seen as activities that generate increasing economies of scale. Given the relatively high population density of the Netherlands, significant economies of scale were envisaged.

In comparison with earlier processes of upscaling in the water sector, the mergers of the late 1990s differed as to the driving forces behind them. While the wave of mergers in the 1970s and 1980s had been driven, if not forced, by government policy (and to some extent resisted by smaller utilities), the mergers in the 1990s were usually supported by the water utilities. Not all companies joined the latest upscaling effort, and at present (2010) the smallest company serves 198,000 connections, whilst the largest one serves 2,340,000 connections (Table 13.1).

Table 13.1 Efficiency gains in the dutch drinking water sector.

Year	Water supplied (million m^3)	Employees (f.t.e.)	Connections per employee
1970	870,766	8138	463
1975	966,048	8328	526
1980	1,011,976	8504	569
1985	1,064,322	8447	637
1990	1,227,000	8422	674
1998	1,171,000	7461	878
2003	1,191,000	5443	1339
2008	1,140,000	4938	1532

Source: VEWIN Drinking Water Industry Statistics, various years.

13.4 BENCHMARKING AND EFFICIENCY

Partly as a result of the increased emphasis on efficiency, Dutch water utilities engaged in benchmarking activities for the first time in the 1980s. For more than three decades this benchmarking scheme was developed during which time its nature and aims changed considerably. What started as a voluntary scheme for performance comparisons became a mandatory scheme for regulating the drinking water sector.

Benchmarking in the Dutch water sector was first undertaken by an association of water supply companies named COCLUWA in the late 1980s. This association introduced a closed financial and technical accounting system, which has led to annual rounds of voluntary performance comparison among COCLUWA members. The system was implemented by the 11 COCLUWA members from 1989 onwards. In the early years, the benchmarking was an inter-company activity whose results were shared by the directors of the participating companies. By the late 1990s, 14 water companies partook in this annual cross-company comparison of performance.

From 1997 onwards, the benchmarking results were made public. This increased emphasis on transparency needs to be seen within the context of the "privatization debate", which reached its peak around that time. Publishing the benchmarking results was a way to show to the general public and the politicians that the utilities were performing well and that privatization and independent regulation were not necessary to improve performance. As such, the decision to publish the benchmarking results was partly a move to sway public opinion and influence the debate. Benchmarking remained voluntary but the role of the Association of Dutch Drinking Water Companies (Vewin) increased. Officially benchmarking served two objectives, namely to provide greater transparency for interested parties, and to provide the water companies with insights on how to improve their processes (Vewin, 2007). The interested parties are the central government, customers, supervisory directors and shareholders, and the drinking water companies themselves (Vewin, 2007). In accordance with the Balanced Score Card framework for selecting four benchmarking perspectives, the water supply companies decided to focus their benchmarking on Water Quality, Service Quality, Environmental Impact and Finance and Efficiency. The voluntary benchmarking exercise has been performed every three years since 1997. Detailed protocols have been developed and refined over the years for each of the four perspectives.

Since 2010, the new Drinking Water Act has stipulated the inclusion of a mandatory benchmark for the drinking water sector, replacing the existing voluntary benchmark. There were three reasons why the policy makers replaced the existing voluntary benchmark with a mandatory system (Van Geel, 2009). Firstly, the government wants to force all companies to participate. The second reason for the government to make the benchmark mandatory is to improve the accuracy and reliability of the data and the quality of comparison. The third reason is that the introduction of mandatory Improvement Plans has given the government an instrument to promote performance improvement.

Interestingly, the mandatory benchmark is also the main barrier to further upscaling of the Dutch water services sector. Benchmarking requires a minimum number of utilities in order to allow useful performance comparison. The current number of water utilities seems to be close to the required minimum. With a limited potential of increasing their scale of operations in the Netherlands, water utilities are increasingly trying to internationalize their activities in order to grow.

13.5 INTERNATIONALIZATION OF THE SECTOR

In the context of the "privatization decade" mentioned above, another development also stimulated the centralization of the Dutch water supply sector in the 1990s. Given that different infrastructure services like the energy market and the railway system were liberalized in the 1990s, the liberalization of the water supply market would have been a "logical" continuation of that policy. The private water supply market, however, was dominated by a few very large multinationals. Therefore, liberalization would have made these companies competitors of Dutch water utilities. In order to be able to compete with these private operators, the Dutch utilities would have had to grow. The multi-utility NUON[2], for example, believed that it was too small to compete with some of the mega-utilities from other countries. "The size of our competitors means that further growth is essential"[3] (NUON, 2001:14).

[2]NUON consisted of an energy and a water supply utility which were merged into a multi-utility in preparation for a completely liberalized market that did not materialize. The merger was subsequently undone, leading to the formation of the VITENS water supply company in 2002. The NUON energy-only company was later absorbed by the Swedish Valsvatten.

[3]Author's own translation.

Another development that has swept through the drinking water sector in recent years is that of getting engaged in international activities, supported by both Dutch Development Aid and International Financing Institutes, on a "not-for-profit" and "not-for-loss" basis. As the "privatization decade" lost momentum, the concept of "Water Operator Partnerships" (WOPs) surfaced as a possible alternative for improving service provision in developing countries. In March 2006, the United Nations Secretary – General's Advisory Board in Water and Sanitation (UNSGAB) led by the ex-prime minister of Japan, Ryutaro Hashimoto, developed a Compendium of Actions entitled the Hashimoto Action Plan (Hashimoto, 2006) under the direct request of the Secretary General Kofi Annan. The objective of this Plan was to put forward a series of actions aimed at boosting the efforts around the world for achieving the MDG targets on water and sanitation, and it recommended WOPs as a useful mechanism to do so. The International Water Association (IWA) defines WOPs as "any formal or informal collaboration or structured partnership aimed at capacity building on a not-for-profit basis. Partnerships can take a multitude of different forms and have various technical, legal and social shapes depending on individual circumstances" (IWA *et al.* 2009).

Quite a few Dutch water utilities have embraced these WOPs in recent years. Apart from pursuing objectives of social corporate responsibility, utilities engage in these WOPs as they are expected to have a positive impact on human resources. That impact is believed to be two-fold. Firstly, the employees engaged in these projects are likely to become both more motivated and learn new approaches and skills in these projects. By becoming more motivated and acquiring new insights and skills, the employees become more productive. Secondly, companies which develop international activities become more attractive to young professionals, thus making it easier for them to attract qualified young professionals in the increasingly competitive Dutch labour market.

As a result of the changes of the past decade, the organizational structure of Dutch utilities has become quite complex. Most utilities are actively engaged in the liberalized market for industrial water supply. In addition, utilities have developed special business units for activities which can be exploited commercially (like laboratory work). These units are organized as special purpose companies under the main utility. Moreover, Dutch utilities, which are officially not allowed to operate abroad, have set up subsidiaries which engage in international activities. The result is that the many public water utilities now consist of a multitude of organizations: a core company (the actual service provider) surrounded by a variety of special purpose for-profit and not-forprofit companies as illustrated by the organizational setup of the largest water utility in the Netherlands Vitens (Figures 13.2 and 13.3).

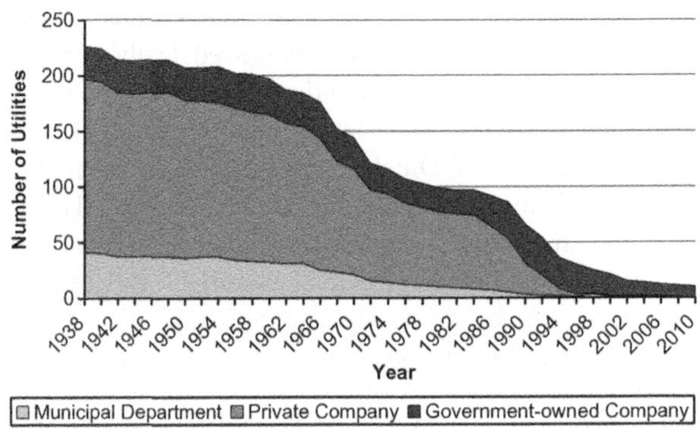

Figure 13.2 Number of companies and organizational form of water supply in the Netherlands, 1938 – 2010.

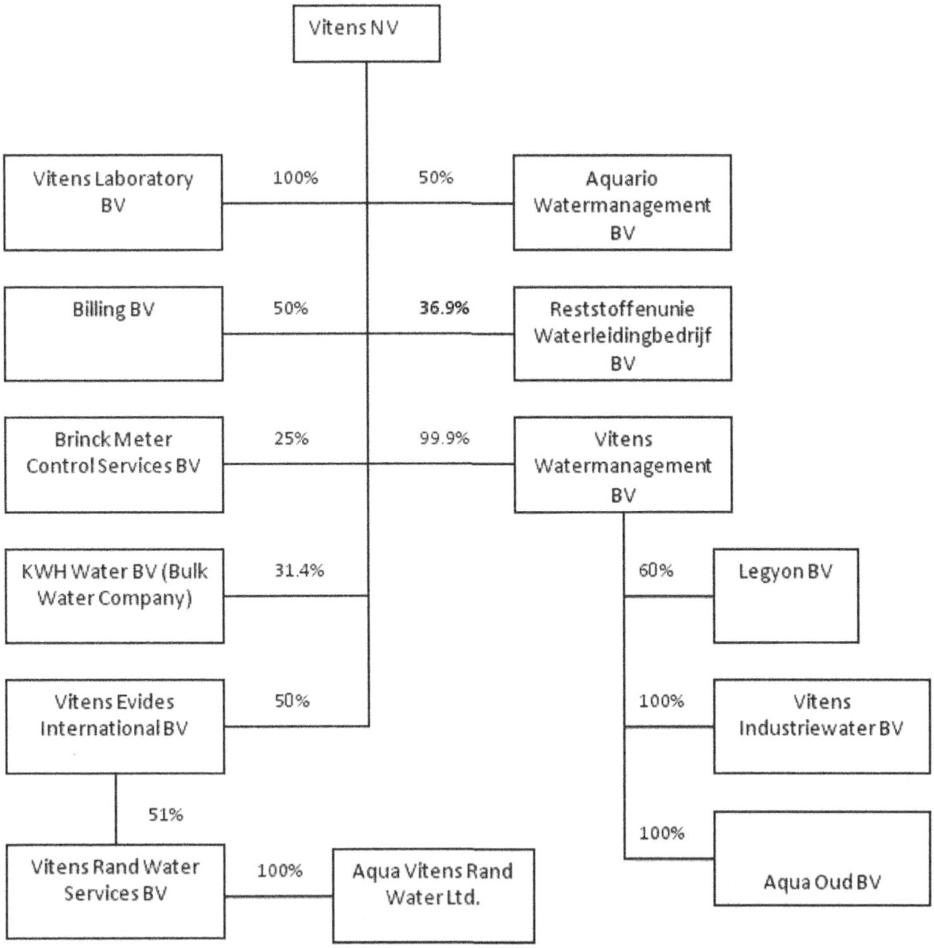

Figure 13.3 Organizational setup of Vitens N.V. and its subsidiaries.
Source: Vitens, 2009; percentages denote ownership of shares; N.V. stands for *Naamloze Vennootschap* which resembles a Limited Liability Company; B.V. stands for *Besloten Vennootschap* which resembles a Private Limited Company.

13.6 CONCLUSION

Over the past 70 years, the Dutch water services sector has been subjected to a continued process of mergers and upscaling of water utilities. Of the 232 utilities in 1938, currently only 10 remain. Although this trend of centralizing service provision has existed for eight decades, the underlying reasons for upscaling have changed considerably over time. Initially, upscaling was an instrument for the central and provincial governments to expand service to remote areas of the country. As service coverage reached 100%, the drivers of upscaling changed. Partly as a result of the (neo-liberal) international climate, efficiency gains became the main objective for water utilities. Mergers were driven and supported by the managements of utilities rather than central and provincial governments. Moreover, upscaling was viewed as a necessity in order to be able to “compete” with large international operators in a liberalized market.

In recent years, as the liberalization of the water supply market has disappeared from the public agenda, and with the limited potential for more mergers in the Dutch water sector, utilities are increasingly pursuing a strategy of diversifying and internationalizing their (commercial) activities. As a result, Dutch water utilities are busy establishing strings of profit-oriented and non-profit oriented subsidiaries.

REFERENCES

Blokland M. (1999). Evolution of the Dutch water supply sector. In: Private Business, Public Owners: Government Shareholdings in Water Enterprises, M. Blokland, O. Braadbaart and K. Schwartz (eds), Ministry of Housing, Spatial Planning and the Environment, The Hague.

Dane P. and Warner J. (1999). Upscaling water supply: the case of Rotterdam. In: Private Business, Public Owners: Government Shareholdings in Water Enterprises, M. Blokland, O. Braadbaart and K. Schwartz (eds), Ministry of Housing, Spatial Planning and the Environment, The Hague.

Dijkgraaf E. and van de Mortel E. G. (1997). *Marktwerking Monopolisten in de Watersector Missen Node Controle* (Market Functioning of Monopolists in the Water Sector Lacks the Necessary Control). *Het Financiële Dagblad*, 9 September 1997.

Hashimoto R. (2006). Hashimoto Action Plan – Compendium of Actions. United Nations Secretary General's Advisory Board on Water and Sanitation, http://www.unsgab.org/docs/HAP_en.pdf. (Retrieved October 14, 2011).

IWA, GWOPA and VEI (2009). Water Operators Partnerships – Building WOPs for Sustainable Development in Water and Sanitation, http://www.unhabitat.org/pmss/listItemDetails.aspx?publicationID=2851. (Retrieved October 10, 2011).

NUON (2001). Jaarverslag 2000. NUON, Arnhem.

Van Geel P. L. B. A. (2009). *Memorie van Toelichting: Nieuwe Bepalingen met Betrekking tot de Productie en Distributie van Drinkwater en de Organisatie van de Openbare Drinkwatervoorziening (Drinkwaterwet)* (Memorandum of Explanation: New Clauses with Respect to the Production and Distribution of Drinkingwater and the Organization of the Public Drinkwater Supply) (Water Supply Act). No. 30895. Ministry of Housing, Spatial Planning and the Environment, The Hague.

Vewin (2007). Reflections on Performance 2006. *Vereninging van Exploitanten van Waterleidingbedrijven* (Dutch Association of Watersupply Companies), Nederland and Accenture, Rijswijk.

Vewin (various years). Watersupply Statistics. *Vereninging van Exploitanten van Waterleidingbedrijven* (Dutch Association of Watersupply Companies), Nederland, Rijswijk.

Vitens (2009). Vitens Annual Report 2008, http://www.vitens.nl/Vitens/Het+bedrijf+Vitens/Publicaties.htm. (Retrieved January 20, 2010).

Chapter 14

Water services management and governance: past lessons for a sustainable future

Ezekiel Nyangeri Nyanchaga

14.1 INTRODUCTION

Kenya is rated among the water scarce countries of the world. In terms of water quality, Kenya's water resources face serious threats from pollution, siltation, reclamation, pesticides, weed attack, and human activities. The development of water supplies in Kenya is appreciable through four distinct periods: pre-colonial (1895–1920), Colony and Protectorate (1920–1963), Independence (1963–1980) and Post-Independence (1980–2000).

African participation in water services management and governance grew exponentially from minimal and negligible – menial work such as digging trenches or ferrying water pipes during the pre-colonial period – to full participation and ownership in the post-independence period. Since water is a basic requirement for human survival, its development is crucial, and various administrative regimes have assigned varying degrees of importance to it. Water development in Kenya, like any other form of development, has faced various challenges.

The general trend, however, has been skewed policy prioritisation to serve those in power, rather than the priority areas; urban areas and administrative centres were provided with water, rural water supply was forfeited. Water sector reforms came in hardy, and in 2002 a new framework to meet the huge challenges to service delivery in the past was introduced. Water quality got minimal attention and the policy governing pollution was delayed until 1972 – even then implementation was not taken seriously. Though the reform process has not been fully completed yet, the new framework has already led to significant improvements in the sector.

14.2 PRE-COLONIAL AND COLONIAL PERIOD (1880–1963)

Until the 1880s, the present day Kenya consisted of unlimited and sedentary pastoral and spatially scarce agricultural economies. There was a universal right to water, but certain water rights were allocated to groups or individuals for specific uses through a social negotiation process (Meinzen & Nkonya, 2005). The inhabitants searched for water from rivers and springs and transported it home in various types of receptacles ranging from clay pots to buckets made from hides.

The construction of Uganda Railways (Figure 14.1) pioneered the development of water supplies in Kenya. Before the start of the building of the Railway in 1896, there was no single piped water supply in Kenya (Mats, 2004). The construction of the Uganda Railways began in Mombasa in 1896, reached Nairobi in 1899 and its destination in Kisumu in 1901 (Marsh & Kingsnorth, 1965) giving the country its first water supplies (Colony and Protectorate of Kenya 1913–1923).

Figure 14.1 Uganda railway near Mombasa c. 1899 (Marsh & Kingsnorth, 1965). Township stations and initial water supply were built along it.

The Hydraulic Branch of the Public Works Department, under the Director of Public Works, undertook management of water by 1901. The HB opened offices in Nairobi in 1902, Kisumu in 1903 and Naivasha, Eldoret and Nyeri in 1910e, (Nyanchaga & Ombongi, 2007). As the British took over the resources of the protectorate, the grip by the African institutions systematically relaxed and the customary role they played was significantly reduced (Lotte, 2006).

The colonial period marked development of urban centres, industries and a centralised government. The colonial government eventually took over the ownership and control of natural resources in the protectorate. In the late 1920s, the state took over urban water supplies from railways (Colony and Protectorate of Kenya, 1930). The Public Works Department developed new township water supplies. The Railways abandoned some water supplies and connected to the new ones (Nyanchaga & Ombongi, 2007).

In this period, the first water legislation was passed. Sikes presented a report called "Modern Water Legislation in 1922 (British East Africa Protectorate, 1923). The draft water legislation was written into a bill in 1928, and Her Majesty assented to the bill in December 1934 (Secretary of State for the Colonies, 1934) and it became the first water legislation on July 1, 1935 (Colony and Protectorate of Kenya, 1935). The legislation covered all surface waters in the state and gave the authority for managing and enforcing the water law to a new government body – the Water Board (Colony and Protectorate of Kenya, 1929). This water law was revised in 1951 and the Minister for Agriculture was given the overall mandate over water development policy. The Minister in charge of water resources was mandated to appoint a Water Undertaker and select an advisory body to assist with policy and the implementation of the legislation (Colony and Protectorate of Kenya, 1929b).

Until the 1940s, there was no written national plan on how to mobilise the country's water resources. Water development in townships was in most cases *ad hoc* and usually undertaken at township level and basis. Each township had its own water development plan cut out (Nyanchaga & Ombongi, 2007). Rural water supplies through settlement schemes in Kenya were the direct result of the decision made by the government to spend 3 million pounds for the reconditioning of African areas and African settlements under the Kenya ten-year development programme. It was not the policy of the African Land Development Board (ALDEV) to maintain the schemes after the initial period of establishment (Waddicar, 1958).

The need for reorganisation led to the establishment of the Commission of Enquiry into the separation of the Roads Branch from the Public Works Department in 1956 described by Herbert J. Manzoni.

He recommended the transfer of the HB to the Department of Agriculture, all supplies in large towns be taken over by the Local Authorities, and that the Ministry of Agriculture (MoA) operate supplies in smaller towns (Cabinet Office, 1957). This model outlived the colonial government by 25 years.

By the time Kenya gained independence in 1963, the governance of the water sector in Kenya appeared to have passed through two stages. The first stage was characterised by the imposition of colonial rule, confiscation of land and water resources, and impartial application of water and land laws – common in the period between 1920 and 1940s. The second stage involved concentrated efforts towards more inclusive legislation and extending water development to the African population. This period that began in the late 1940s saw the revision of the Water Ordinance of 1929 in 1951, and the implementation of the Dixey scheme. By the end of the colonial period, three parties were involved in water supplies provision: the Ministry of Works (MoW), the Water Development Department and the Local Authorities which resulted in duplication of duties and hence inefficiency. The water supplied was polluted since the Water Resources Authority, Public Health Authorities and the Water Apportionment Board mandated with the task lacked the support of appropriate legislation. Standards were crucial since without them it was difficult to ascertain cases of pollution or institute prosecution (Colony and Protectorate of Kenya, 1931–1970).

14.3 INDEPENDENCE PERIOD (1963–1980)

After independence in 1963, the new government used five-year development plans to harness the rapid development of the republic. The first development plan from 1964–70 was mainly a carry-over from the colonial period whose focus was economic growth (Ochieng, 1995). Water development was declared important for the economy, and priority was given to schemes that were expected to be financially self-sustaining, such as water services for the municipalities (Government of Kenya, 1964).

Many actors who had previously been involved in water governance proceeded with their work, while others entered the scene as a result of independence and reshaped the landscape of water development. In the1970s, water became a highly prioritised area in for the government. According to WHO, urban areas had almost universal coverage meeting the international standard by 1972 (WHO, 1972). However, as the economy slowed down in the 1970s and the financial stability of urban water supplies began to dwindle, the ministry operating urban supplies could no longer recover capital costs. The Water Act (Cap, 372) did not clearly spell out what constituted a violation. Therefore, it was deemed that any new legislation on pollution control should be made part of the Water Act either by introducing an amendment to it or by reformulating the pollution rules of the Act (Republic of Kenya, 1963–1972). Consequently, in 1972, the Water Department released an Interim Report on the water Pollution Policy for Kenya which stated that the national goal to provide water for all in Kenya by the year 2000 had to go hand in hand with sewerage so as not to destroy the water sources through pollution. The report noted that 85% of the population at the time depended on untreated water However, enforcement was affected by the lack of a serious attitude toward pollution, the multiplicity of regulations, cumbersome legal procedures, and lack of trained enforcement personnel. At the same time, there was no single organisation fully responsible for the administration and enforcement of laws (Ministry of Agriculture, 1972).

In November 1974, a full-fledged Ministry of Water Development was created to handle water affairs. A subsequent WHO review mission observed that the financial situation was not sustainable and would result in deterioration of infrastructure. Expansion of urban water services was also compromised by the failure of the government's housing policy. By 1979, it was clear that the government's goal of "water for all by the year 2000" was not going to be met. Deterioration of the financial situation continued, and by 1984 the urban supplies under MOWD were only able to recover operation and maintenance costs. Public service levels deteriorated over the 30 years following independence (Nyanchaga & Nilsson, 2008). However, the government and the international organisations continued to support provision of high quantity and quality water supply.

14.4 POST-INDEPENDENCE PERIOD (1980–2000) AND LATER DEVELOPMENTS

The post-independence period marked another milestone in Kenya's water sector development. In the 1980s, the Government of Kenya issued long-range coverage targets for the International Drinking Water Supply and Sanitation Decade. Targets were originally given up to the year 1990 but were extended later to 2000. The improved water supply coverage of 100 per cent in urban areas (approx. 5 million people) and 75 per cent (approx. 16 million people) in rural areas was to be met in 1990 (Hukka, Katko & Seppala, 1992).

During this period the government experienced financial constraints with regard to meeting the set targets by the year 1990, hence the revision of the target year to 2000. It also became clear that, on its own, the Government could not deliver water and sanitation services to all Kenyans by the year 2000. Therefore, attention turned to finding ways of involving other players: local authorities, communities, NGOs and the private sector in the provision of water services in place of the Government (Hukka, Katko & Seppala, 1992). On the other hand, in the middle of the 1980s the GTZ started the Urban Water and Sanitation Management (UWASAM) Project in the Ministry of Local Government, to assist water undertaking municipalities to improve provision of water and sanitation services.

In 1988, the Government established the National Water Conservation and Pipeline Corporation (NWCPC), as a state corporation to take over the management of Government operated water supply systems that could be run on a commercial basis. The corporation was made an implementing body responsible for relatively large scale water development projects. Its formation was an attempt to reform the sector by getting large viable water supplies schemes to operate outside the direct control of the Ministry of Water Development. However, the NWCPC was not able to meet the desired goal as the cost of developing the scheme could not be covered by water revenues. Thus, it continued to depend on the Central Government for capital development. Revenues could not cover operation and maintenance costs or capital costs. Therefore, large municipalities, community water groups, non-governmental organisations (NGOs), co-operatives and private water services providers were brought in to complement the water supply system.

People not served under any of the above arrangements did not have a systematic water service, and had to collect water directly from a watercourse or some other water source on a daily basis. In 1991, the Ministry of Water Development and SIDA agreed to undertake a Delineation Study in an effort to enable the actors of the sector to cope with the anticipated developments in the sector by identifying their future roles, functions and responsibilities. The National Water Master Plan Study was also completed in July,

1992. The two documents made recommendations about reforms in the water and sanitation sector, but did not set clear-cut policies that the government could have bought into for immediate action.

The current situation with water and sanitation services in Kenya is far from satisfactory. Service coverage and service level are inadequate both in urban and rural areas. The performance of existing water supply systems is poor due to various constraints. Recently, Kenya has set challenging development targets to improve water and sanitation services. A major effort is the water sector reform process started to reorganise the sector to achieve the development targets.

Some water sector actors are mainly concerned with policies, others with implementation, operation and maintenance while some have multiple roles, which has led to contradictions in the sector. Other actors indirectly involved in the water sector include the Ministry of Agriculture (MOA), the Ministry of Health (MOH), Regional Development Authorities, Kenya Wildlife Service (KWS), various non-governmental organisations (NGOs), self-help and community groups, as well as private companies

The primary role of the water and sanitation sector in the *Kenya Vision 2030* (the Vision) is to ensure availability and access to water and improved sanitation for all. The rationale for the inclusion of the water sector in strategic goals was not just social impetus and the human right to water. Indeed, the Vision identifies the "specific linkages between water and the Vision's economic and social pillars (Government of Kenya, 2007). It was contemplated that water and sanitation services demand would likely reach unprecedented levels as the country develops and achieves its growth targets in industry, urbanisation and population growth. In order to meet the demand for water and sanitation services, eight areas covering scarcity, security, quality of water and sanitation services, infrastructure development, general management and monitoring have to be addressed.

14.5 WATER SECTOR REFORM INITIATIVES AND THE BOTTLENECKS

The major challenge in water development has been how to increase coverage of services and sustain good quality at the same (Republic of Kenya, 2008). Today, the water and sanitation sector faces major challenges related to the years of adjustments in O&M and infrastructural development (Table 14.1). Population increase (Figure 14.2), patterns of land use, low investment in the water and sanitation sector, poor maintenance and non-replacement of obsolete equipment and failure to install new assets are some of the major challenges the country's water sector has undergone over the years.

In the light of the current water reform initiative, very few reform initiatives are new. Most had been studied and recommended but not implemented. The concept of reform is nothing new in the water sector as contemporary knowledge seems to indicate. The earliest major reforms in Kenya were carried out in the pre-colonial period when traditional water governance ceded the ownership of water governance to the colonial government. More inclusive reforms took place after independence following the launching of the policy document "Sessional Paper No. 10 of 1965" which directed the Government's policy towards priority areas for the African population identified as poverty, illiteracy and diseases. The policy required that the core infrastructure for economic and social activity be in Government hands. Accordingly, the Government was engaged in all productive activities, including the provision of water and sanitation services. The Sessional Paper No. 1 of 1986 provided another platform for major reforms in Kenya and the water and sanitation sector.

Rural-to-urban migration and overall population growth among other demographic dynamics have led to stagnation in the spread of water service coverage in the country. Nairobi, for instance, has consumed the largest share of water supply funds for more than three decades in expansions, extensions of the existing systems, repairs and other related works. This has been at the expense of small towns and rural areas.

That is why the population in other cities (Figure 14.3) and major towns remains poorly covered and the rural population neglected.

Table 14.1 Main phases of water and sanitation development in Kenya starting in the 1960s (Nyanchaga and Ombongi, 2007).

Period	Descriptive trend	Objectives ("ideal")	Implications ("real life")
1960s–mid 1970s	Social services and poverty alleviation	Free basic water and sanitation services by the government	Unrealistic targets: coverage remained low
mid 1970s–early 1980s	Supply-driven approach & self-help initiatives	Self-help initiatives ensuring community participation	Many self-help schemes were taken over by the government
early 1980s–1990s	Decentralisation and commercialisation	Local authorities and NWCPC get increasing responsibilities	De-concentration instead of decentralisation
mid 1990s–Present	Towards a more decentralised and delegating policy	Community water supply. Emerging policy: delineation and clarification of roles.	Declining government resources triggered policy changes
Present–Future	Policy and reform implementation. Private sector participation.	Delegation. Re-regulation. IWRM. Resource mobilisation. Capacity building. Sanitation concerns.	Reform impacts still to be seen

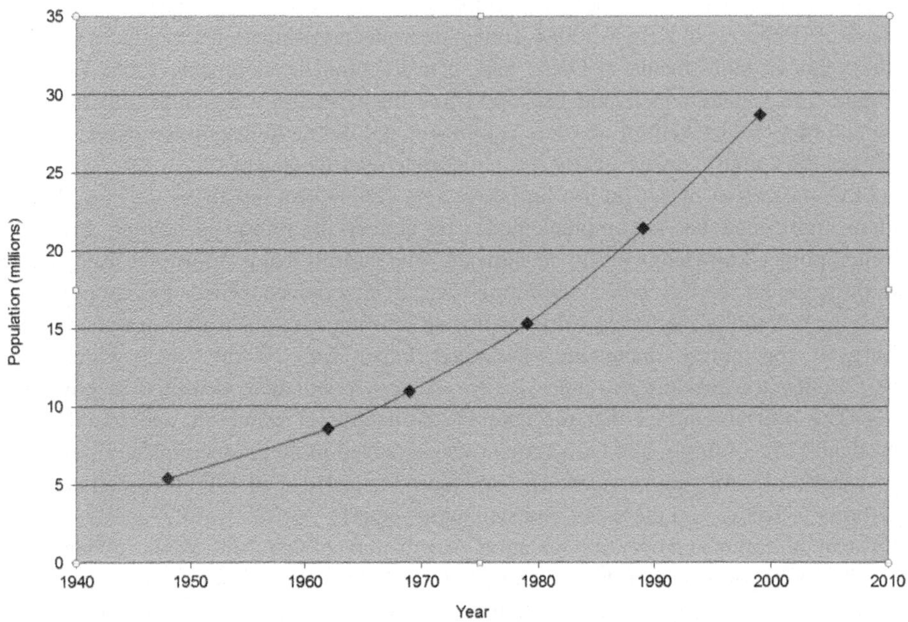

Figure 14.2 Population census graph 1948–1999.

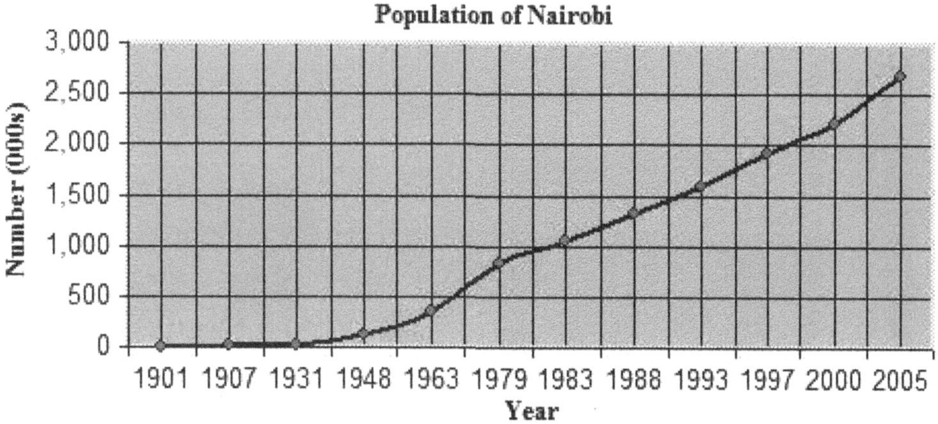

Figure 14.3 Nairobi population graph between 1901 and 2005.
Source: Republic of Kenya, 2002a.

The thrust for reform was given impetus by the recognition that water supply and sanitation strategy must be made a national priority. That required understanding the factors that necessitated reform in the water sector in Kenya. Prior to 1999, the national water sector policy was derived from the five year national development plans, recommendations of the first (1980) and second (1992) National Water Master Plan Studies, and national policy papers on other sectors and issues. The national water policy in Kenya is outlined in Sessional Paper No. 1 of 1999 on National Policy on Water Resources Management and Development. The policy states that the Government's role would be redefined to focus on policy and coordination while direct service provision would be left to municipalities, the private sector and communities. The policy also states that the Water Act Chapter 372 Laws of Kenya would be reviewed and updated, and attention would be paid to the transfer of water facilities.

The policy justified handing over, arguing that ownership of a water facility encourages proper operation, maintenance and ownership. The policy stated that the Government could hand over urban water systems to autonomous departments within local authorities and rural water supplies to communities. This was also consistent with international resolutions including those of the Mar del Plata Conference of 1977 and the Rio de Janeiro Earth Summit (1992) that led to agenda 21. At the national level, the new approach and management of water affairs was meant to support the aspirations of the National Environmental Action Plan (NEAP) and consequently the Environment Management and Coordination Act (EMCA), 1999, which was developed at the same time as the national water policy.

The Water Act 2002, as previously noted, marked the beginning of a much-needed overhaul of the water and sanitation sector in Kenya. The Act's shift in policy centered on four themes of reform that would: (i) separate management of water resources from provision of water services; (ii) separate policy making from operations and regulation of the water sector; (iii) decentralise functions to lower-level entities; and (iv) provide a framework for the involvement of civil society in both management of resources and provision of services (Muma, 2007) (Figure 14.4).

Based pm recommendations made at the World Summit on Sustainable Development in Johannesburg, the MWI together with the WRMA began work to produce the Integrated Water Resources Management (IWRM) and the Water Efficiency (WE) Plan for Kenya. These were published jointly in August 2009. Kenya is challenged by the fact that it has to harmonise its development priorities with international standards. The adoption of international standards interferes with national priorities, renders them irrelevant and results in wastage of resources. Other global challenges include the emerging requirements of human rights, and fulfillment of the Millennium Development Goals.

Figure 14.4 Public health officer Rachuonyo (Kenya) demonstrating hand washing processes using a tippy tap (Kunguru, 2009). *Source*: Kenya water supply and sanitation, 2009.

The 2002–2008 National Development Plan (Republic of Kenya 2002a) indicated that of the 142 gazetted urban areas in Kenya, only 30% have sewerage systems. According to Nippon Koei (1998), the majority of sewers constructed in the 1950s, '60s and '70s have never been inspected and are in poor structural condition.

In addition to water sector reform as per the Water Act 2002, in 2007 the country unveiled *Kenya Vision 2030*, a long-term national development plan that would serve as a "vehicle for accelerating transformation of the country into a rapidly industrialising middle-income nation by the year 2030 (Government of Kenya, 2007) Vision 2030 is faced with the difficulty of predicting the future due to frequent in changes in demand for water, population growth, and the inability to understand and predict with any degree of confidence the rate of constitutional implementation. On the other hand, assumptions made in the promulgation of these policies and visions, for example that the economy will grow 10 percent annually in the next 25 years, are influenced not only by internal implementation challenges but also by various external and uncontrollable factors. This leads to failure during implementation.

The new constitution establishes a new governance framework that has radically changed the governance and management of public affairs in Kenya. In harmonising the provisions of the new constitution for the water sector, it is imperative that due recognition is given to the key provisions in relation to the role of the state, the rights of citizens and the institutional mechanisms for the enforcement of basic rights. Reorganisation of the water sector – and the diminishing role of the government in actual service provision – is expected to reduce political interference in the operation of water systems. However, it is probable that politicians and other individuals with vested interests will continue to interfere in water management at the regional or local level.

The challenges of addressing the concerns raised in the New Constitution are in implementation. As in the case of previous studies and propositions, recommendations have been shelved for years rendering them irrelevant for the prevailing conditions. Undoubtedly, despite the importance of various studies and their recommendations, implementation has always very difficul. The gap between policy and implemented reality has been wide and discernible as in the case of the reallocation of water responsibilities which was recommended in 1992 and implemented in 1999 and the agreement on the formation of regional water bodies recommended in the 1960s and implemented in the 1990s.

14.6 DISCUSSION AND CONCLUSIONS

The development of Kenya's water sector in the last century provides a basis for planning sustainable development. All along, the major objective has been attainment of more coverage and better quality while all efforts, though diverse in outcomes, were geared towards one objective. Attainment of this objective has been a rough path challenged by increased population, migration patterns, management and implementation patterns, lack of resources, and environmental issues. Although reforms have been initiated along the development process, current reform initiatives are intended to comprehensively address both emergent and recurrent challenges.

The proposed reform process is well planned and reflects an appropriate interpretation of the new policy and strategy objectives.It is assumed that various stages and decisions along the process will be successfully implemented according to the plans. However, there is no proper assessment of the uncertainties, risks and alternative approaches for situations where implementation does not proceed as planned. A culture of shelving recommendations only to later implement them, when their relevance has diminished, or just delaying the benefits of implementing them should be avoided. There is a need to dissect and consequently identify the core issues leading to the failure of the water for all by the year 2000 strategy in order to succeed with the water as human right concept entrenched in the new constitution and Kenya Vision 2030.

There are currently challenges in sustainability of institutions running dilapidated infrastructure, lack of adequate sector investment policies, coordination and communication, sector governance, human resources development, and fragmented capacity development strategies. Hence, the weakness of the sectoral visions has been that they did not adequately consider the overall social and economic development in Kenya, which is a prerequisite for a successful water sector reform. Adequate financial resources need to be available and effectively managed in order to maintain the operational works. Inspection, repairs and maintenance need to be consistently carried out in order to avoid the mistakes of the past.

Reorganisation of the departments responsible for water supply was first recommended by the Williams Bransby Report in 1907. In the 1920s the Railways handed over the urban water supplies to the state, and in 1957 Herbert Manzoni recommended reorganisation of the water sector, followed in 1970 by the WHO and in 2002 by the Water Act 2002. In 2011, the amendment of the constitution provided a new challenge to the structural organisation of water sector by placing water resources under the control of the land commission and water service management under the Ministry of Water. All these reform initiatives were intended to restructure the institutions responsible for water supply. They suggested both decentralisation and centralisation leaving the sector disoriented without achieving much. It is crucial to determine the extent to which the reorganisation of the responsible departments is effective, and whether these reorganisation Were necessary.

A development plan including specific measurable indicators for various objectives is needed for equitable water services and the orientation of their development. Water development should have tangible milestones to establish the level of progress towards achievement of a particular goal. The plan should be executable independent of the government of the day, and the civil servants working within the ministry responsible for water development should be empowered to carry on the development work on schedule even if the political will were not available.

Recognising the international contribution to and requirements of the local water development scene, an approach that accommodates the dynamism of the sector is important in order to meet both the local and global needs of the water sector. A national water council where all actors can deliberate and operationalise the water development agenda should be established.

REFERENCES

British East Africa Protectorate (1923). The Water Ordinance, Crown Advocates Department. Kenya National Archives, Ref: AG/43/87, Nairobi, Kenya.

Cabinet Office (1957). Council of Ministers. Kenya National Archives, ACW/32/30. 26th March 1957.

Colony and Protectorate of Kenya (1913–1923). Muthaiga Water Supply. Kenya National Archives, Ref: AG/43/103, Nairobi, Kenya.

Colony and Protectorate of Kenya (1929). Official Gazette 31 Dec 1929, Water Ordinance 1929. Kenya National Archives, Ref: AG/43/79, Nairobi, Kenya.

Colony and Protectorate of Kenya (1929b). Official Gazette, 31 December 1929 Water Ordinance 1929. Kenya National Archives, Ref: AG/43/79, Nairobi, Kenya.

Colony and Protectorate of Kenya (1930). Report for 1929 His Majesty's Stationery Office, London Ref: AG/43/79, Nairobi, Kenya.

Colony and Protectorate of Kenya (1931–1970). Human Disaster and Natural Calamities, Water Pollution. Kenya National Archives Ref: BY/64/1, Nairobi, Kenya.

Colony and Protectorate of Kenya (1935). The Official Gazette, 25 June 1935. Kenya National Archives, Ref: AG/43/79, Nairobi, Kenya.

Government of Kenya (1964). Development Plan 1964–1970. Government Printers, Nairobi, Kenya.

Government of Kenya (2007). Kenya Vision 2030: A Globally Competitive and Prosperous Kenya, Oct 2007, Government of Kenya, Kenya, 116.

Hukka J., Katko T. and Seppala O. (1992). Kenya Water Sector – A Position Paper. Tampere University of California Press, Berkeley. USA.

Kenya Water Supply and Sanitation (2009). Hygiene and Sanitation Promotion in Schools. Lake Victoria South Water Services Board. Ministry of Water and Irrigation, Kenya.

Koei N. (1998). The Aftercare Study on the National Water Master Plan in the Republic of Kenya. Final Report, Executive Summary. Japan International Cooperation Agency, Ministry of Water Resources, Republic of Kenya. November 1998.

Lotte H. (2006). Moving the Maasai, A Colonial Misadventure. Palgrave Macmillan, in Association with St Anthony's College, Oxford.

Marsh Z. A. and Kingsnorth G. W. (1965). An Introduction to the History of East Africa. 3rd edn, Cambridge University Press, British Institutes of History and Archaeology in East Africa.

Mats W. (2004). Towards a Historical Geography of Intensive Farming in Eastern Africa. Islands of intensive agriculture in East Africa. The British Institute in East Africa, 10 Carlton House Terrace, London, SWIY 5AH.

Meinzen D. R. and Nkonya L. (2005), Understanding legal pluralism in water rights: lessons from Africa and Asia, paper presented at International workshop on 'African Water Laws: Plural Legislative Frameworks for Rural Water Management in Africa', 26–28 January 2005, Johannesburg, South Africa. http://www.nri.org/waterlaw/AWLworkshop/MEINZEN-DICK-R.pdf/ (assessed 01 November 2006)

Ministry of Agriculture (1972). Interim Report on Water Pollution Policy for Kenya. Water Department. Kenya National Archives, Ref: RP/3/205.

Muma A. (2007). Kenya's New Water Law: an Analysis of the Implications of Kenya's Water Act, 2002, for the Rural Poor, International Workshop on *'African Water Laws: Plural Legislative Frameworks for Rural Water Management in Africa,'* Johannesburg, South Africa, 26–28 January 2005.

Nyanchaga E. N. and Nilsson D. (2008). Pipes and politics: a century of change and continuity in Kenya Urban Water Supply. Journal of Modern African Studies, **461**(2008), 133–158. Cambridge University Press, United Kingdom.

Nyanchaga E. N. and Ombongi K. S. (2007). History of Water Supply and Sanitation in Kenya, 1895–2002. Environmental History of Water, Chapter 21. IWA Publishing, Alliance House, 12 Caxton Street, London, UK, 2007.

Ochieng W. R. (1995). Structural and political changes. In: Decolonisation and Independence in Kenya 1940–93, B. A. Ogot and W. R. Ochieng (eds), James Currey Ltd, London, pp. 83–109.

Republic of Kenya (1963–1972). Water Pollution Control Policy General. Water Development Department. Kenya National Archives, Ref: RP/3/205.

Republic of Kenya (2002a). National Development Plan 2002–2008. Effective Management for Sustainable Economic Growth and Poverty Reduction. Government Printers, Nairobi, Kenya.

Republic of Kenya (2008). Background and Status of Reforms in the Water and Sanitation Sector in Kenya. Ministry of Water and Irrigation, Nairobi, Kenya. February 2008.

Secretary of State for the Colonies (1934). Governor of Kenya, Dispatch dated 1934-12-3. Kenya National Archives, Ref: AG/43/80, Nairobi, Kenya.

Waddicar H. (1958). Settlement Schemes in Kenya. Kenya National Archives, BY/6/2. 28th October 1958.

WHO (1972). Sectoral Study and National Programming for community and rural water supply. Sewerage a Dam water pollution. Report NO. 2 Recommendations' on National Programme for Community Water Supplies Development. Brazzaville WHO.

Chapter 15

Prisoners of a paradigm? What can water sector donors learn from history?

David Nilsson

15.1 INTRODUCTION

In 1969 the Kenyan government declared an ambitious goal; to provide all citizens with water by the year 2000 (Government of Kenya, 1969). Numerous governments in developing countries have declared similar intentions since then, through the many UN-led development fora and processes: the Mar del Plata conference 1977; the UN Water Decade 1980–89; the 1990 New Delhi conference; the Millennium Development Goals in 2000. These processes have all had support from the large Western donors, who have invested billions of dollars in improving water and sanitation in low-income countries in order to reduce poverty world-wide (Black, 1998). Important gains have been made, especially in South and East Asia. But in Africa, south of the Sahara, there are still huge deficiencies in access to water and sanitation, and the MDG target on water will not likely be met there (WHO and UNICEF, 2010). Kenya has even seen a downward trend, not least in the cities. While official figures indicated more or less universal coverage of piped water in Kenyan towns in the 1960s, today coverage of piped water in urban areas is reported to be 47% (WHO, 1963; WASREB, 2011).

Donors have been part of the development processes in Africa for fifty years. They have a responsibility for the outcomes of these processes, and it is high time donors take a look at their own role in the history of water development in developing countries. This paper hopes to demonstrate how donors could start learning from history and their own experience. Using a case study from Nairobi, Kenya, I will look at the role of the World Bank and how the poor have been included in water provision.

In Nairobi, around 63% of the 3.1 million inhabitants are served by the water utility (WASREB, 2011). According to a recent survey carried out by the French Institute for Research in Africa (IFRA) and the Global Water Operators Partnership Alliance (GWOPA), there are huge differences within the city in terms of water consumption as well as what consumers pay for water. In some parts of the city, consumers enjoy piped water well in excess of 200 litres per capita per day while in the informal settlements – the "slums"–the per capita daily consumption is as low as 13 litres. The richest 10% of the population consume around 45% of the water in Nairobi, while the poorest 45% get around 15% (Ledant 2011; Ledant forthcoming). The price of water paid by consumers in these areas is often more than 20 times the tariff for piped water (ibid; Castro 2009). According to the IFRA/GWOPA study, water consumption in Nairobi is highly dependent on housing type and income class: high per capita consumption occurs in the high-income areas west of the city centre (see Fig. 15.1).

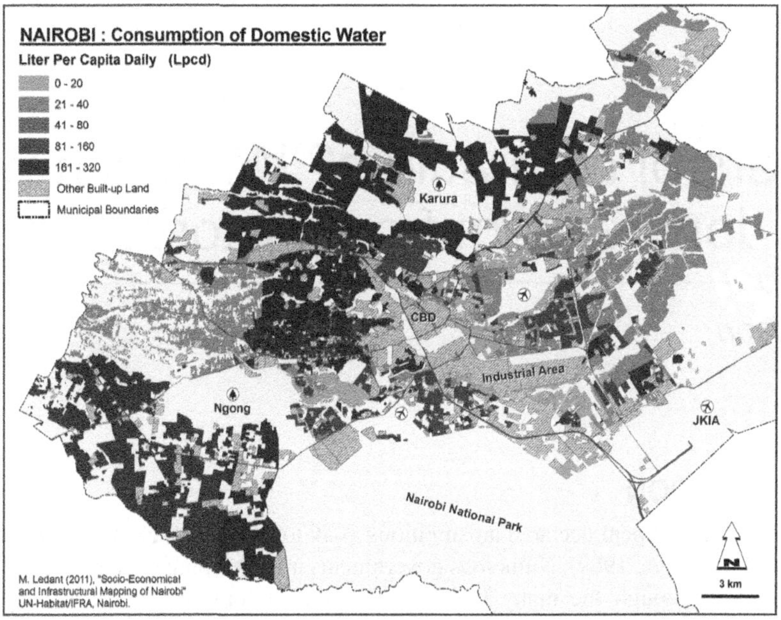

Figure 15.1 Daily water consumption in Nairobi. Darker shading indicates higher per capita consumption. *Source*: From Ledant 2011.

Kenya adopted a new Constitution in 2010 which awards every citizen the right to safe water in adequate quantities. In the light of the constitution, and the international human rights debate, the inequalities in the public water supply in Nairobi are hardly justifiable. How did this unequal situation develop? What key decisions led to such a segregated service provision regime? In the following, I will describe the historical trajectory of water supply and demand in Nairobi to start unpacking these questions.

15.2 WATERING A THIRSTY CITY

15.2.1 A water demand history of Nairobi

In 1899, the new railway from the coast to Uganda had progressed almost all the way to the escarpment of the Rift Valley. At this place, known as "Enkare Nairobi"; a "place of cold water", a major railway depot was built (Obudho and Obudho, 1992). Over and above its strategic location before the descent into the Rift, there was enough fresh water to cover the needs of the railway depot (White *et al.* 1948). While the available water was deemed adequate in 1899, there has since been a constant struggle to keep supply in line with the increasing demand. Total demand in 1907 was estimated at around 1200 m^3/day (British East Africa Protectorate 1907). Current demand – including losses and water for non-domestic uses – has been estimated by various consultants at 417,000 to 670,000 m^3/day (EGIS 2011). Total demand does not only follow growth of the population but is also a result of a changing social structure and consumption behaviour. When designing the systems, engineers have to estimate how much water each user is going to consume, the "per capita design demand". Over time, design demand has changed in response to social changes. Below, I will analyse in more detail how the concept of design demand has evolved over time in Nairobi as part of local engineering practice.

In 1907 the design demand for domestic consumption in Nairobi was 18 gallons per head, or 80 litres per capita per day (lpd). At the time no distinction was made between the different consumer groups (British East Africa Protectorate, 1907; White *et al.* 1948). However, the design demand used in Nairobi was soon to be stratified according to race. In 1934, design demand was 225 lpd for Europeans, 135 lpd for Indians and 90–68 lpd for Africans, depending on education level (Colony and Protectorate of Kenya, 1934). In 1945, the same practice was followed except that Africans were allowed 112 litres per person (Colony and Protectorate of Kenya, 1945). After independence in 1963, design demand in Nairobi was based on income class rather than race (see Table 15.1).

Table 15.1 Design demand for domestic water in 1969.

	Design demand (lpd) - domestic
High income	270
Average Urban	158
Low income	68
"Squatters"	18

Source: Nairobi City Council 1969.

Table 15.2 Design demand for domestic water as used in various contemporary study reports.

	MWI (2005)	**Seureca (2007)**	**Egis (2011)**
High class housing	250	275	240
Middle class	150	175	144
Low class	75	75	72
Low class without connection	20	20	36

Source: Egis 2011.

Today, the general design demands of the sector are found in the Ministry of Water and Irrigation (MWI) Practice Manual from 2005. Other recent design reports for Nairobi have used the Practice Manual figures modifying them slightly, as shown in Table 15.2.

For analytical purposes, the different consumer categories over time can be grouped into four broad social groups. The emerging picture shows that the design demands per capita for domestic consumption in Nairobi have not changed very much since the 1930s on the whole (see Figure 15.2). The definitions of the different consumer categories have varied over time, from racial to an income-based stratification. But the design demand figures seem to have warranted very little or no discussion in recent times. Interviews with key actors in 2012 indicate that design demand norms are generally taken as "a given" in the water sector (Owoucha; Gakubia; Mwangi, personal communication 2012). However, the design demand estimate has a huge impact. A high design demand for the high-end consumers makes it more difficult to re-allocate water between users, since there appears to be no "free" water to re-allocate. Considering demand an exogenous factor – "a given" – may thus increase total demand. In effect, design demand is not simply an engineering issue but also a normative statement with political and ethical implications.

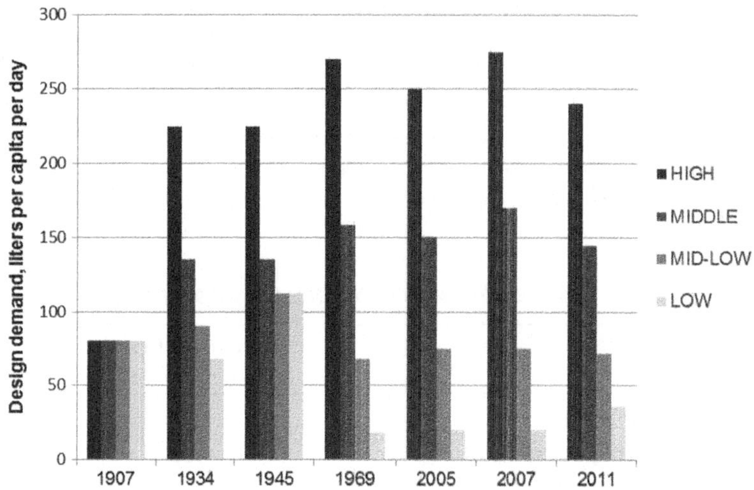

Figure 15.2 Design demands (litres per capita per day) for domestic use by different consumer classes in Nairobi between 1907 and 2011. Before 1969 classes were based on race, after that on income groups.

15.3 A SUPPLY-SIDE HISTORY OF NAIROBI'S WATER SUPPLY

15.3.1 The "periodical bogey of water shortage"

Kenya is a water scarce country. The sanitation expert sent out by the Colonial Office in 1907 – George Bransby Williams – reported that the available supply in Nairobi was not more than 40 litres per capita per day (British East Africa Protectorate, 1907). The city saw perennial water shortages throughout the 1920s. In 1926, there was not enough water to even supply the fire hydrants, and strict conservation measures were developed (Nairobi Municipal Council, 1926; Smart, 1950). In 1928, the Council decided to construct a new supply from the Ruiru River, north-west of the town (Smart, 1950). The aim was to allow Nairobi to "be free from the periodical bogey of water shortage", as expressed by the Mayor of Nairobi, Lady Delamere, at the opening ceremony in 1938 (East African Standard, Nov 17, 1938). But in early 1946, the "bogey" was back to haunt Nairobi. As an emergency measure, the Nairobi dam was built on the outskirts of the city to store water from the Ngong River (Colony and Protectorate of Kenya, 1948). As a more long-term solution, the Sasumua dam was built on the Chania River, completed in 1956 (Colony of Kenya, 1956). Water was now being piped a distance of some 56 km and the inhabitants of Nairobi received, on average, around 140 litres per person per day (ibid, Smart, 1950). Nairobi thus entered the 1960s relatively well equipped with access to water.

15.3.2 Post-colonial water: enter the World Bank

"*In the urban areas, the problem is primarily one of ensuring the adequacy of supplies for the rapidly rising population. However, this involves major investments.*" Republic of Kenya (1974).

The capacity of the Sasumua dam was augmented in 1968 (EGIS, 2011). Still, Nairobi City Council calculated that capacity would soon be outstripped by growing demand. As indicated by the quote above, the leaders of post-independence Kenya also feared the periodical water shortage and wanted to continuously invest to keep this "bogey" at bay. In 1969 the Council turned to the World Bank as a source for finance (Nairobi City Council, 1969). In the following 25 years, the Nairobi City Council would carry out an array of large projects to improve water supply with the support of the World Bank: the First (1970–1977), Second (1978–1984) and Third (1985–1994) Nairobi Water Supply projects; the Nairobi Sites and Services Project (1975–1982) and; the Second Urban Project (1978–1986).

The three consecutive Nairobi Water Supply Projects struggled to keep in pace with rising demand (see Fig. 15.3). The new projects tapped into a new water source at the Chania River to the east and downstream from Sasumua. The projects created a whole new water lifeline to the east of the city. It was complemented with transmission lines and new mains in the eastern part of the city. The three projects were concluded in 1994 with the construction of a large dam at Thika, some 50 km north of Nairobi (SOGREAH, 2005). Today, about 85% of Nairobi's water is supplied from the new eastern system (EGIS, 2011).

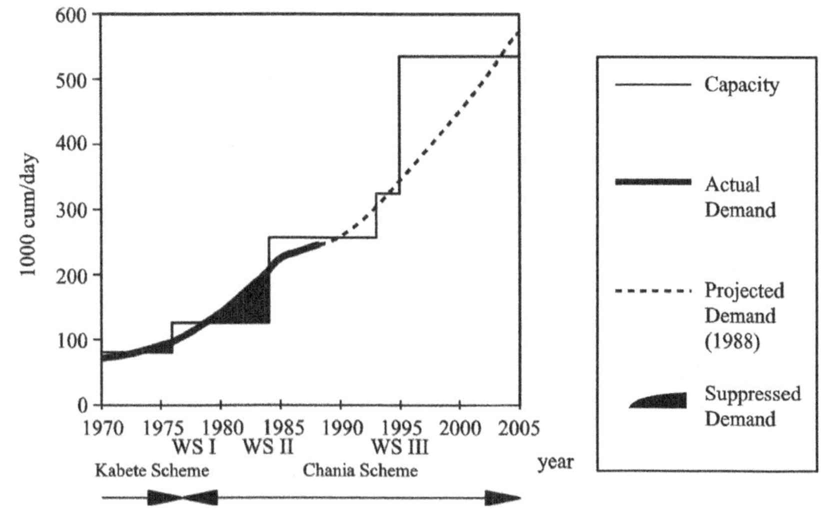

Figure 15.3 The progressive expansion of raw water supply during Nairobi water supply projects 1–3. *Source*: World Bank 1996.

At present, the maximum raw water supply is around 520,000 m^3/day not counting private boreholes. However, the actual supply available after production and transmission losses is around 350,000 m^3/day, compared to the estimated demand of more than 600,000 m^3/day (SOGREAH, 2005; EGIS, 2011). Since the turn of the Millennium, water shortages have been almost permanent with regular rationing taking place (SOGREAH, 2005; AWSB, 2010; Ledant, 2011). Already in the 1990s, plans had been drawn up – again with World Bank support–to further augment Nairobi's water supply in the future. Even far-off sources north of the Thika dam were to be tapped through a system of collector tunnels (ADB, 1998). In 2011, the Athi Water Services Board carried out a feasibility study for this "Northern Collector Scheme", to bring the total raw water supply up to 1.3 Mm^3/day by 2035 (EGIS, 2011).

15.4 DISTRIBUTION AND WATER EQUITY

There are around 400,000 connections in Nairobi serving some 63% of the population directly through individual connections; the remainder receive water through yard-taps, kiosks or water vendors (WASREB, 2011; EGIS, 2011). As shown by numerous studies, consumption is lower and price higher in areas that do not have access to the piped network (UN-Habitat, 2003; Bousquet, 2010; Nilsson, 2011). Equity is therefore closely associated with the layout of the distribution system.

15.4.1 Distribution in planned areas

During the colonial period, the network reflected the priorities of the ruling colonial elite. A minimum of services were provided in African housing areas, and the Nairobi City Council spent no more than a few percent of its budget on these areas (East Africa Royal Commission, 1955; Zwanenberg and King, 1975). After independence, a major expansion of the distribution mains took place in the 1970s and 1980s through the First Nairobi Water Supply projects (World Bank, 1996). The secondary distribution network was also expanded through the two Urban Projects. In the Third Nairobi Water Supply Project in the 1990s, some distribution mains were built, mainly in the outlying areas to the east and south (ADB, 1998).

Today, the distribution network is deemed inadequate in most parts of town. The water operator is presently dealing with many problems of service unreliability and low pressure, exacerbated by leaks and illegal connections. Moreover, it is today not possible to measure and control how much water flows into different pressure zones due to the lack of bulk meters and dysfunctional valves (Ruhiu; Muguna, pers. comm. 2012). Some planned low-income areas with high-rise buildings are hard hit by the low pressure, since there is not enough pressure to serve the upper floors (Muguna, pers. comm. 2012).

15.4.2 Distribution in unplanned areas

For many years, informal areas were neglected by the formal service provider, as they were considered illegal. The official policy of the authorities in the 1960s and 1970 was typically to forcibly remove "squatters", or just ignore these areas with regards to service provision (Bousquet, 2010; Nilsson, 2011). From the end of the 1970s, the government's vigilance against informal settlements gradually relaxed, to which donors may have contributed. Notably, in 1977 the World Bank demanded the inclusion of a component for water services in informal settlements during the preparations of the Second Nairobi Water Supply Project (Erkmen, 1977). The First and Second Nairobi Water Supply Projects constructed water kiosks in the informal areas, while the two Urban Projects focussed on upgrading informal areas into planned and serviced neighbourhoods (World Bank, 1996). The Third Nairobi Water Supply Project initially failed to include service provision in informal areas, but a component for more water kiosks in Kibera was added towards the end of the project, in the late 1990s (ADB, 1998).

15.5 INSTITUTIONAL AND ORGANISATIONAL CHANGES

Water services were provided by the Nairobi City Council (NCC) between 1923 and 2004. In terms of financial sustainability, the NCC for a long time collected enough revenue to generate a substantial surplus (City Council of Nairobi, 1973; Nilsson and Nyangeri, 2008). However, by the end of the 1990s, a large billing backlog had developed. Many accounts had huge arrears, there were many illegal connections, and un-accounted for water had soared to 52%. NCC was reported to be financially unsustainable and was only kept from insolvency through subsidies from the Government (Halcrow, 2001). By the mid-1990s, the situation had become completely unsustainable, and water sector reforms were initiated – supported by the World Bank and others – to deal with poor sector performance. The new Water Act of 2002 meant that all urban service provision is by commercial operators. In 2003, the Nairobi City Council formed a wholly owned limited company, the Nairobi City Water and Sewerage Company, NCWSC (Castro, 2009).

While sector reforms have resulted in improved economic performance, service provision is still visibly highly unequal. The institutional framework provides few incentives for NCWSC to improve services in the low-income areas since there is no monitoring of equity at the utility level (Nilsson, 2011). Moreover, as long as the supply of raw water is limited, supplying more water to the poor means selling less water to the rich. The high-end consumers pay more per unit due to the raising block tariff structure. Hence, this provides a commercial disincentive for the utility to provide more water to the poor areas (see also Ledant, forthcoming).

15.6 CONCLUSIONS

The World Bank has no doubt been the most influential donor with respect to Nairobi's water supply. Although the Bank promoted service provision to low-income groups already in the 1970s, in terms of equity, there is very little to show for all the money and time spent. More water has come to Nairobi, but very little has "trickled down" to the poor. What general lessons could water sector donors learn from this story?

15.6.1 Prisoners of a paradigm

The history told here indicates that water development in Nairobi is confined within what the historian of technology Giovanni Dosi (1988) calls a "technological paradigm". This paradigm is based on large-scale piped water supply and European standards of consumption, with middle and high-income consumers as the main target group. A technological paradigm does not consist only of technology. It also includes routines, norms and knowledge about how things should be done; what Dosi refers to as "a set of heuristics". Design demand is an example of a norm within this paradigm. The present design demand structure was established in the 1930s and has essentially remained unchanged. There are few indications of the prevailing technological paradigm ever having been questioned, and the main focus of the system-builders remains to supply increasing amounts of water in the fight against "the bogey of water shortage".

By making financial resources available for the successive expansions of water supply, the World Bank has indirectly supported the preservation of a technological paradigm which has mostly benefited the middle- and high-income earners. It is true that the World Bank also has promoted and financed distribution in low-income areas. But these interventions have neither had the intended effect, nor been sufficient to reduce the overall inequality in provision. The World Bank is, after all, a bank, and the supply-oriented technological paradigm has provided a vehicle for the Bank to fulfil its lending mandate. This does not prove that there is an inherent goal conflict between re-distributional and demand-oriented approaches, on the one hand, and the lending mandate of the Bank, on the other. But the history of water in Nairobi indicates that the World Bank and other donors have become prisoners of a supply-oriented development logic where large investments are favoured and where equity issues tend to attract less attention. An important question is therefore what the key organisations in the sector have learned from their experiences over time, and how that learning can facilitate change.

15.6.2 Self-sustaining feedback

The World Bank has a well renowned evaluations department which regularly evaluates its operations. Furthermore, the Water and Sanitation Programme (WSP) was formed in 1978 to improve structural learning regarding water provision for the poor within the World Bank Group, among other tasks (Black, 1998). In 1996, the Operations Evaluations Department made a large evaluation of the Bank's three Water Supply Projects and two Urban Projects, spanning 20 years. The evaluation found that on most accounts, the water projects had fulfilled their objectives. The number of water kiosks had been increased by 150, and overall water supply had multiplied. However, it was also noted that the water kiosks were subject to cartel-like organisations that hiked the prices and thus did not deliver the intended pro-poor benefits. As to the urban projects, the upgrade of housing and services led to gentrification of the population, thus simply moving the poor somewhere else (World Bank, 1996). Furthermore, other studies have shown that the late inclusion of an informal settlements component in the Third Nairobi Water Supply Project was marred by financial problems and was not carried through to completion (Katui-Katua and McGranahan, 2002). But the failure to provide affordable services to the poor did not alter the Bank's overall rating of the water projects: the outcomes were all rated "satisfactory" (World Bank, 1996). This lends strength to the argument that the low-income components of the projects were more of an "add-on" nature while large-scale lending was the real core of the operations.

What did the Bank learn from twenty years of support and USD 120 million of lending? The main recommendation based on the 1996 evaluation had to do with the poor sustainability of the projects. A main lesson – according to the Bank's evaluation report – was to make sure there was sufficient institutional capacity in place before venturing into more large-scale lending activities to Nairobi City. In effect, this meant changing the institutional and organisational setup to better match the large-scale technological paradigm. This recommendation fits well into the pattern of prescriptive policies for sector reforms having been propagated by the World Bank Group from around 1994 (Black, 1998; Bayliss, 2003). Consequently, the World Bank has since been a strong supporter of sector reforms and, among other things, financed the restructuring of the Nairobi water utility in 2004–2007 (World Bank, 2008).

Another lesson learnt was that community-based approaches and water kiosks for low-income areas failed because water cartels took over. The World Bank currently plans to finance individual connections in slums to be operated by the water company (Mwangi, pers comm 2012). The idea is thus to circumvent the cartels and extend the large-scale supply-oriented paradigm to all consumers. The Bank is at the same time preparing the ground for yet another large-scale lending operation to bring more water to Nairobi; the USD 1 billion Northern Collector Scheme (EGIS, 2011).

In essence, the learning mechanisms involved in the Nairobi case were programmed to sustain the paradigm; not to question it. Past decisions about engineering standards have come to contain a political meaning of inequality and segregation but these standards are not questioned. The lessons that currently are being fed back into the donor machinery are about how to make the context work for the technology. This is what is currently meant by "sector reform". But should not the lessons be about how to change the technology to work for the context?

15.6.3 Epilogue: history as a development laboratory

There are no magic bullets for sustainable and equitable water supply in Nairobi or elsewhere. But if poverty reduction and improved equity really are the goals of donors, then learning and policy development should be put into its longer historical context, and questions asked about what really makes an impact for poor people. It is time that donors start looking back, and not just keep on looking into the haze of tomorrow with a preconceived "set of heuristics". History research provides an under-utilised source of knowledge for development actors. It is a laboratory for analysing and understanding what enables or hinders a more equitable society. The study of history is our chance to understand the future.

Acknowledgements

Some of the data presented in this paper was collected in collaboration with the Institut Francaise de Recherche en Afrique (IFRA) in Nairobi and the Global Water Operators Partnership Alliance (GWOPA). The author wishes to thank both organisations.

REFERENCES

Literature

Bayliss K. (2003). Utility privatisation in Sub-Saharan Africa: a case study of water. *Journal of Modern African Studies*, **41**(4), 507–531.

Black M. (1998). Learning What Works. A 20 year Retrospective View on International Water and Sanitation Cooperation. World Bank WSP, Washington, D.C.

Bousquet A. (2010). Water and the poor in Nairobi: from water apartheid to urban fragmentation. The case of Kibera. In: Nairobi Today. The Paradox of a Fragmented City, H. Charton-Bigot and D. Rodriguez-Torres (eds), IFRA, Nairobi.

Dosi G. (1988). The nature of the innovative process. In: Technical Change and Economic Theory, G. Dosi, C. Freeman, R.R. Nelson, G. Silverberg and L.L.G. Soete (eds), Pinters Publishers, London, pp. 221–238.

Katui-Katua M. and McGranahan G. (2002). Public Private Partnerships and the Poor: Small Enterprises and Water Provision in Kibera, Nairobi. WEDC, Loughborough.

Ledant M. (2011). Socio-economic and infrastructural mapping and analysis on Nairobi: GRUBS city level. Technical Report. IFRA, Nairobi.

Ledant M. (forthcoming 2012). GRUBS City Level Nairobi Phase II. In-Depth Analysis. IFRA and GWOPA, Nairobi.

Nilsson D. and Nyangeri E. N. (2008). Pipes and politics: a century of change and continuity in Kenyan urban water supply. *Journal of Modern African Studies*, **46**(1), 133–158.

Nilsson D. (2011). Pipes, Progress, and Poverty. Social and Technological Change in Urban Water Provision in Kenya and Uganda 1895–2010. Doctoral Thesis in History of Technology, Royal Institute of Technology, Stockholm.

Obudho R. A. and Obudho R. A. (1992). The colonial urban development through space and time 1895–1963. In: An Economic History of Kenya, W. R. Ochieng and R. M. Maxon (eds), East African Educational Publishers Ltd, Nairobi, pp. 147–158.

UN-Habitat (2003). Water and Sanitation in the World's Cities. Earthscan, London.

WHO and UNICEF (2010). Progress on Sanitation and Drinking-Water. 2010 update. WHO Press, Geneva.

World Bank (1996). Kenya Impact Evaluation Report. Development of Housing, Water Supply and Sanitation in Nairobi. Operations Evaluations Department. Washington, D.C.

van Zwanenberg R. M. A. and King A. (1975). An Economic History of Kenya and Uganda 1800–1970. Macmillan Press Ltd, London.

Other written sources

African Development Bank, ADB (1998). Completion Report, Third Nairobi Water Supply Project, Kenya. ADB.

Athi Water Services Board, AWSB (2010). Request for proposals, Consultancy services for Nairobi city water distribution network modification. AWSB/WaSSIP/Comp 1/CS-29/10. AWSB, Nairobi.

British East Africa Protectorate (1907). Report on the sanitation of Nairobi and Report on the Townships of Naivasha, Nakuru and Kisumu, by G. Bransby Williams. Kenya National Archives: GP 363.7 BRI.

Castro V. (2009). Case Study on Institutionalization of Water and Sanitation Services to Informal Settlements. World Bank, WSP–AF, Nairobi.

City Council of Nairobi (1973). Nairobi Sewerage and Groundwater Survey. Organization, management, finance and legislation report. Part II, 30 September 1973. SWECO, Stockholm – Nairobi. Kenya National Archives: GP 628.2058 NAI.

Colony and Protectorate of Kenya (1934). Report, Ruiru Water Supply Scheme. 13 June 1934. Kenya National Archives: RN/1/88.

Colony and Protectorate of Kenya (1945). Report on Augmentation of Water Supply for the Nairobi Municipal Area. By F. E. Kanthack, Consulting Engineer. Government Printer, Nairobi. Kenya National Archives: BV/19/295.

Colony and Protectorate of Kenya (1948). Annual Report for the Year 1946. His Majesty's Stationary Office, London. Nordic Africa Institute Library, Uppsala.

Colony of Kenya (1956). Report of the Commissioner for Local Government for the year 1956. Kenya National Archives: K 352. 06762 0005 KEN.

Constitution of Kenya (2010). Laws of Kenya, National Council for Law Reporting, Kenya.

East Africa Royal Commission 1953–1955 (1955). Report. Her Majesty's Stationery Office, London, reprinted 1961. Nordic Africa Institute Library, Uppsala.

East African Standard, newsarticle November 17, 1938. Kenya National Archives: BV 19/295.

EGIS BCEOM (2011). Feasibility Study and master plan for developing new water sources for Nairobi and satellite towns. Water Sources Options Review Report. Report to Athi Water Services Board 6 Sept, 2011.

Erkmen Esref. (World Bank) in a letter to Mr Moche, managing director of NCC - WSD, dated 28 Feb 1977. Kenya National Archives: RN/6/44.

Government of Kenya (1969). Development Plan 1970–1974. Government Printer, Nairobi.

Halcrow Group (2001). Study of options for Private sector participation in the Water and Sanitation Services of Nairobi. PSP Options Report, September 2001. Report to Government of Kenya.

Nairobi City Council (1969). 1952–1969, Preliminary Report on the Work of the Water and Sewerage Study Group, Feb 1969. Kenya National Archives: RN/6162.

Nairobi Municipal Council (1926). Report of the Nairobi Water Committee, 12 November 1926. Kenya National Archives: RN/4/63.

Republic of Kenya (1974). Development Plan 1974–1978. Part 1. Government Printer, Nairobi, Kenya.

Republic of Kenya (2008). Background and Status of Reforms in the Water and Sanitation Sector in Kenya. Final Report, February 2008. Ministry of Water and Irrigation, Kenya.

Smart J. A. (1950). Jubilee History of Nairobi. Kenya National Archive: GP 307.76 SMA.

SOGREAH (2005). NWSEPIP Component 1: Diagnostic of the current situation & definition of an emergency physical investment programme. Report to Nairobi Water Services Board (AWSB).

Water Services Regulatory Board, WASREB (2011). Impact Report No. 4. WASREB, Nairobi.

White L. W. T., Silberman L. and Anderson P. R. (1948). Nairobi Master Plan for a Colonial Capital. Her Majesty's Stationery Office, London. Kenya National Archives: 711WHI.

World Health Organization, WHO (1963). The Organization for Water Development in Kenya. Report of a WHO Consultant Team. Kenya National Archives: BY/138/43.

World Bank (2008). Implementation Completion and Results Report. Nairobi Water and Sewerage Institutional Restructuring Project. Report No: ICR0000835, World Bank Africa Region.

Interviews

Gakubia, Robert (WASREB). pers comm 2012-01-23.

Muguna, Nehemia (Nairobi City Water and Sewerage Company) pers comm. 2012-01-26.

Mwangi, Patrick (World Bank/WSP) pers comm 2012-01-25.

Owuocha, Kenneth (Athi Water Services Board), pers comm 2012-01-18.

Ruhiu, John (Nairobi City Water and Sewerage Company) pers comm 2012-01-26.

Chapter 16

Reforming water services in the Helsinki Metropolitan Area

Petri S. Juuti and Riikka P. Rajala

16.1 INTRODUCTION

Water services in the Helsinki Metropolitan Area went through radical change at the beginning of 2010. Until then, there were four autonomous municipal water utilities: Helsinki Water, Espoo Water, Vantaa Water and Kauniainen Water. Political pressure forced these utilities to merge. The new utility – HSY Helsinki Region Environmental Services Authority – is the biggest in the Nordic countries, providing water and wastewater services to about 900,000 people, as well as other environmental services like solid waste disposal. This new utility is bigger than the next three biggest Finnish utilities together. There were many reasons for this merger, but political pressure was probably the major one. The pressure was created by the current ideology of mergers of municipalities which still continue to be promoted by the Finnish Government. For example, in 2009 a total of 99 municipalities merged into 32. Between 2001 and 2011 the number of municipalities in Finland decreased by 112. Over half of them (67) disappeared in 2009. Finland had a total of 336 municipalities in 2012. Municipal mergers also affect water services, since most Finnish water utilities are owned by municipalities. In fact, there are ongoing negotiations at various levels to merge all the municipalities in the Helsinki Metropolitan Area. The factors and processes that led to that will be discussed below.

The Helsinki Region Environmental Services Authority (HSY) is a regional authority providing environmental services for residents and companies in the Helsinki Metropolitan Area (Figure 16.1). Its principal duties comprise water and waste management as well as provision of regional information services. HSY began its operation on 1 Jan. 2010. The joint municipal authority operates the water works of Espoo, Helsinki, Kauniainen and Vantaa as well as the waste management and regional and environmental information services of the Helsinki Metropolitan Area Council, YTV. HSY has a staff of about 750. About 570, or more than 70 per cent, of them are water professionals.

HSY produces drinking water at its two water treatment plants in Helsinki and one in Espoo. Municipal and industrial wastewater is treated in the Viikinmäki Wastewater Treatment Plant in Helsinki and the Suomenoja Wastewater Treatment Plant in Espoo. HSY builds and repairs the water pipe and sewer network with a length of approximately 7600 km. The treatment results of the Viikinmäki plant are top quality by international standards. They remove 95 per cent of the phosphorus and about 90 per cent of the nitrogen from wastewater. HSY is a member of the International Water Association (IWA), an international umbrella organisation operating in the water supply and sewerage sector, and of the Finnish Water and Waste Water Works Association (FIWA). HSY Water participates in the operations of the Baltic Sea Action Group (BSAG). HSY serves its customers in its offices in Helsinki, Espoo and Vantaa. For the time being, all customers are served in offices located in their home towns. The Espoo office also serves customers living in Kauniainen. HSY is the biggest actor in the field in Finland. Indeed, it is huge compared to other Finnish actors like Tampere, Turku and Oulu (Figure 16.2).

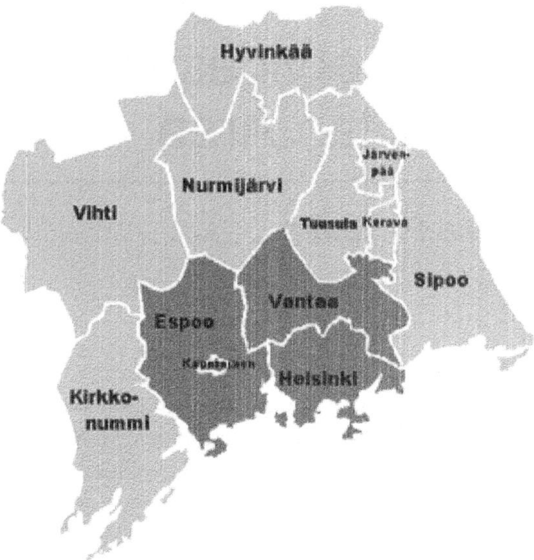

Figure 16.1 Map of the Helsinki Metropolitan Area.

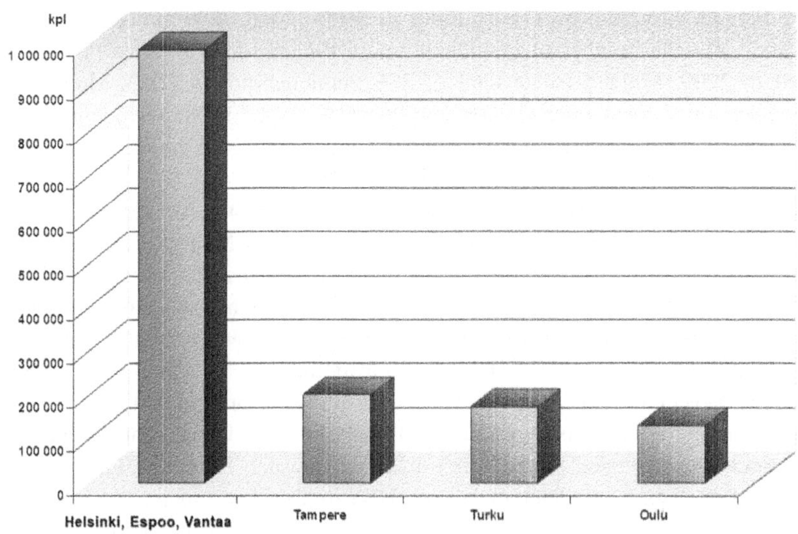

Figure 16.2 Number of inhabitants connected to the network in the Helsinki Metropolitan Area (Helsinki, Espoo & Vantaa), Tampere, Turku and Oulu in 2007.

16.2 MERGING OF SMALLER UTILITIES

In Finland, some small municipal utilities are willing to merge their water services with those of larger ones or sometimes with regional companies. However, their willingness depends largely on existing conditions and their needs. Several intended mergers were never realised, one of the sticking points being the value of the assets. The managers of a small municipality have to be experts in all matters; also those beyond their field of expertise. "*Fortunately midwifery does not lie in my sphere of responsibilities*", said the civil servant of one small municipality. It can be a relief for small municipalities, whose water acquisition, distribution and cleaning have been operated without a proper plan, to hand over responsibility for water supply and sewerage to people who are recognised experts in their field. On the other hand, some smaller utility experts stress the importance of all-round skills: all staff members know what to do when rapid repair is needed. In the case of bulk supply companies or federations, a small number of experts operate the utility according to the principles of lean management. Thus, it is obviously not a question of size only.

On the other hand, mergers of completely separate businesses, such as energy and water works, haves not created synergy benefits as argued by their promoters. Many concerns have been expressed by the water sector, especially about what happens to the spirit of the Water Services Act when such mergers take place. The metropolitan area of Finland (PKS) decided to serve an example for the country. There, water supply and sewerage (WSS), waste management and regional and environmental information services were merged by politicians from the beginning of 2010. The birth of the Helsinki Region Environmental Services Authority (HSY) and the first year of its operation were by no means painless and transparent. The change was not managed properly – for example, the views of the staffs of water works were totally overlooked. However, water supply and sewerage service did not suffer due to the skilled and experienced staff of water works. Functioning WSS was given priority, whereby the customers did not even notice the change.

16.3 EARLY EXPERIMENTS WITH MERGING UTILITIES IN HELSINKI

The fusion that took place in 2010 was not the first try at merging water utilities in the Helsinki Metropolitan Area. Different options had been discussed earlier but had not led to action. Yet, contract-based co-operation with the neighboring water utilities had been fruitful over the years. For example, Helsinki had delivered water to Tapiola, Espoo since 1953. Due to increasing water consumption and problems with the quality of the raw water of Vantaa River, a 120 km rock tunnel from Lake Päijänne was constructed. Helsinki paid 60 per cent of the planning and research costs of the tunnel, while Espoo and Vantaa both contributed 20 per cent. The Päijänne Tunnel was taken into use in 1982. Ultimately, the tunnel had over 10 cities or communities as owners and users. Bilateral co-operation expanded as well: Espoo started to sell water to Kirkkonummi in 1977 and to the western part of Vantaa in 1982. (Water works annual reports, 1970–1982; Erkola, 1982:45) Espoo treated wastewaters from some parts of Vantaa and Helsinki also received some wastewaters from Vantaa (Figures 16.3 and 16.4).

A so-called "tripartite agreement" had entered into to co-ordinate water services in the Helsinki Metropolitan Area already in 1965. The committee members were the water utility managers of Helsinki, Vantaa and Espoo and a representative of the Helsinki Metropolitan Area Council, YTV. Their report and proposal, "Development of Co-operation for Water and Sewage Services in the Metropolitan area" was submitted in November 1988. (Vesi- ja viemärilaitostoiminnan yhteistyön kehittäminen pääkaupunkiseudulla, 1988; Länsiväylä, 5.3.1989).

Figure 16.3 Head office of Viikinmäki Waste Water Treatment Plant in Helsinki.
Source: Photograph Rajala, 2007.

Figure 16.4 Viikinmäki Waste Water Treatment Plant is built underground. The figure shows biological post-filtration.
Source: Photograph Rajala, 2007.

The proposal had two major options for further development in the region. Co-operation could be carried out under a joint organisation or on a contractual basis. Three options for a joint organisation were investigated:

(a) A federation of municipalities
(b) The Helsinki Metropolitan Area Council, YTV
(c) A joint-stock company.

Of these three options the joint-stock company was considered the most feasible. The idea was that decisions about the water network and water prices would be made by communities themselves. The new joint-stock company would be responsible for the main networks and renovations, treatment plants, water utilities and financing. (Länsiväylä, 5.3.1989).

16.4 PROPOSAL 1991: WATER AND SEWERAGE SYSTEM OF THE METROPOLITAN AREA

A so-called contract of co-operation for water services had been signed for 20 years in 1974 and it was going to run out in the early 1990s. A new contract was needed and a decision had to be made whether to renew the contract or merge the water utilities of the metropolitan area. A so-called tripartite agreement further developed the idea of displayed in 1988, and in January 1991 another proposal was made which also included Kauniainen. This later proposal was further thought and it included even draft of contract to sign. But time was running out. The decision had to be made in 1994 at the latest. (Kolmisopimustoimikunnan esitys, 9.1.1991.; Lindberg, 24.4.2008.)

Yet, the 1991 proposal was not accepted by the Espoo City Board. The others decided to wait to see what Espoo would do, because it was known that Espoo would be the hardest to get to agree to the deal. After Espoo said no, everybody else could take a breath of relief. Since the proposal to merge the water utilities was rejected, they decided to proceed on a contractual basis. Espoo and Vantaa signed a contract to pay for the water they received from Helsinki water plants. And Kauniainen and Espoo renewed their bilateral contract in 1993 originally signed in 1975. (Lindberg, 24.4.2008; Länsiväylä, 26.9.1993)

16.5 DISCUSSION

16.5.1 New public management

The described plans and other discussions clearly indicate the arrival of the ideas of New Public Management (NPM) to the Helsinki Metropolitan Area. NPM is a philosophy followed by many governments since the 1980s modernising the public sector. It is a broad and very complex term used to describe the wave of public sector reforms throughout the world since the 1980s. The main hypothesis of the NPM reform wave is that more market orientation in the public sector will lead to greater cost-efficiency for governments without negative side effects on other objectives and considerations. In Finland the NMP reform started at the turn of the 1990s.

According to Vinnari (2008), it meant among others that the responsibility for providing key services to inhabitants, such as water supply and sewerage, remained with the municipalities, but the actual service production could be delegated to another public or private actor or company. Municipalities were also allowed to take into use the financial management models and business bookkeeping of the private sector and to make long term economic plans and budgets. Specifically in water supply and sewerage, the reform included the reorganisation of the former water departments and sewerage system departments, institutions or units as the most independent units and separation of the accounting systems of these new units from that of the municipality.

After the rejection of the 1991 proposal, the idea of a multi-sectoral municipal water company or merger of the water utilities faded from public discussion until 1998 an article on co-operation concerning the water network between Helsinki and Vantaa appeared in *Helsingin Sanomat*. It was considered a canard – there were no official discussions going on about merging the utilities. The parties were satisfied with the prevailing situation. (Lindberg, 24.4.2008).

16.5.2 The metropolis policy and decision making

The so-called Paras Project gave birth to the entire process of creating HSY (Table 16.1). In the metropolitan area, the national Paras Project meant in practice the setting up of 14 teams to think about how to improve co-operation between municipalities. The project was subject to heavy political pressure, because the metropolitan area was expected to serve as an example for the rest of Finland. One team concentrated on water supply and sewerage and how their interests could be combined. That team, called the VVR Group, completed its work on 23 Nov. 2007. It proposed that after reaching an agreement on the establishment of a joint organisation, preparations would be continued based on a limited liability company and a joint authority entreprise model. These models would then be revised and the functionally and economically best solution chosen. (Juuti & Rajala, 2011).

Table 16.1 2007 water services statistics for Helsinki Metropolitan Area and Stockholm.

	Helsinki Metropolitan Area	**Stockholm Vatten**
Turnover (MEUR)	179	131
Pumped water (million m^3)		
– to own operating	89	106
– to other municipalities	1	27
Total	90	133
Number of inhabitants connected to the network	985,000	887,000
Length of network (km)		
– water supply network	2800	2200
– sewer network	2588	1896
– stormwater network	1750	1202
Total	7138	5298
Staff	564	590

However, in practice, all work done by the VVR Group was totally ignored, and the idea of merging all water works under the Helsinki Metropolitan Area Council was brought up. It appeared that the Helsinki Metropolitan Area Council was at risk of becoming an idle organisation because the aim was also to divest it of control over traffic arrangements. The PKS co-ordination group accepted on 12 Feb. 2008 the Helsinki Metropolitan Area Council plan presented by the mayors. In March 2008 they started to prepare the merger of the water works under the Helsinki Metropolitan Area Council. Helsinki Metropolitan Area Council, YTV, was founded in 1970 to deal with the joint affairs of Helsinki, Espoo, Kauniainen and Vantaa. Originally it served as an advisory body until its function changed in the 1980s when it took over traffic and waste management.

While expecting the official decisions, the water utilities started to prepare themselves for merging into a single large water utility. They were convinced that a political decision was about to be made, and that it would be advantageous for the water works to be prepared for it. Espoo and Vantaa initially resisted the merger, but in joint negotiations between their water utilities in late winter and spring of 2008 mutual understanding was reached. It was thought better to seek the best possible result in the negotiations.

The matter was settled permanently at the crucial meeting of Espoo City Council on 19 Jan. 2009, which launched the co-operation with the Helsinki Metropolitan Area Council. Management of the change as such was difficult, because all the preparation work by the municipal officials over the years had totally rejected the particular chosen model. The experiences and knowledge about the good and bad sides of earlier merger efforts were available but totally ignored during the preparation by municipal officials. The Helsinki Metropolitan Area Council model had already been found the least popular, and the worst suited for water services. Yet, the Helsinki Metropolitan Area Council model was jammed through by the politicians.

Although the decision on the merger of the water works under the Helsinki Metropolitan Area Council had been made, there were also other issues that had an impact on the final result. A team preparing the law worked believing that the whole Helsinki Metropolitan Area Council would be abolished. Accordingly, the water works and the Helsinki Metropolitan Area Council would merge under the existing Local Government Act and become a voluntary joint municipal authority. And that is what happened, which ended the story of the Helsinki Metropolitan Area Council. The name of the organisation changed to HSY, but it is also highly likely that the old models of the Helsinki Metropolitan Area Council also transferred to the new organisation as an unnecessary burden.

Instead of allowing the water supply and sewerage sector, clearly the biggest actor of the new entity, to lead the development of the new joint authority, it was once again silenced. Top-down politics was involved in the establishment of the new organisation in many ways and on several levels. For example, the post of the executive director of water supply and sewerage was subject to external application contrary to other posts. The matter was thrashed out in newspapers (e.g. HS 8 Oct. 2009 and 9 Oct. 2009). According to the tenets of good governance, the views of sector experts should also have been considered seriously instead of top-down command-and-control. In other words, participation of sector professionals as well as other stakeholders should have been promoted.

16.6 CONCLUSIONS

Based on all the public material and documents, it is obvious that the real reason for the merger was metropolis policy and politics – not the need to develop water services. The politics were particularly visible in the filling of the post of the executive director of water supply and sewerage and in the choices of persons for the top level management of HSY. Water services are the biggest branch of HSY by all indicators and have to have a clear role. This is because of their role is based on the Water Services Act. According to the Act, the revenue from WSS also has to be used to cover the needs of WSS. That requires a strong and independent WSS organisation regardless of the administrative arrangements.

Amidst all the political play, disappointments and lack of confidence, the merger may still be a possibility for the biggest actor in the environmental field in the Nordic Countries. Especially if the huge amount of knowledge and know-how of the WSS organisation is utilised and developed further under stable conditions and in the positive spirit of the Water Act (Table 16.2). The future will show whether the present organisational form is final or whether for example, the EU will order a reorganisation.

Table 16.2 Lessons to learn and way forward in the water services in the Helsinki capital city area.

Lessons to learn	Way forward
Nobody was given the mandate to lead the process. The head of Helsingin Vesi was given the task directly by the City of Helsink, but he was not given the power or mandate to act as a leader. Such a situation put everyone on their guard.	A clear leader with full leader's rights is needed in the process of change from the beginning.
Clear targets were not set for the work. Each side advocated its own objectives which lead to conflicts.	Clear defined targets must be set for the work. The objectives must be explained to everyone.
The process was not transparent. What were the real motivations behind the merge?	The process must be executed transparently. The motivations are to be explained.
The path dependence of the development of water works was totally eliminated. The merger alternatives, a joint-stock company or a federation of municipalities were passed over and the Helsinki Metropolitan Area Council (YTV) model was chosen.	The path dependence of the past, the present and the future has to be taken into consideration.
The generated savings did not move the charges of WSS. The spirit of the water act is not clearly reflected in the HSY model.	The letter and spirit of the water law are to be followed.
The changes which are started outside the branch based on exterior motivations will usually fail.	The need for change will be appreciated if it comes from inside the branch. The reaction to a need for change from the outside will be critical.
The long and time-consuming process frustrated the staff from the management to the plumber.	A clear time limit is to be set for the change.
Fear of weakening of owner control affected the choice of the organisation. It was feared that the city and citizens could no longer have a say in water supply and sewerage or possible investments.	In the formulation of common rules all parties are to be listened to. Questions or fears are not to be belittled.
The spirit of togetherness within the branch made discussions about difficult issues and agreement possible.	It must be understood that water supply and sewerage is a public good. The water works need not need compete against each other but should operate for the common good.
The environmental sector and environmental services including WSS are used as a political tool in reforming the municipality structure of the metropolitan area. Especially the voice of WSS is not heard within the new, massive organization called HSY.	WSS is a matter of life and death. It should not be a political tool. The own activity of the branch is the best method to be recognized. The appreciation of the whole HSY will rise by appreciating the WSS.

Source: Juuti and Rajala, 2011.

It will also interesting to see what happens to the cities of Helsinki Metropolitan Area. It has been argued that they should be merged into a single mega city on Finnish scale.

REFERENCES

Erkola P. (eds) (1982). *Päijänne-tunneli.* (Päijänne Water Tunnel) Pääkaupunkiseudun Vesi Oy.

Juuti P. and Rajala R. (2011). Veden vai metropolipolitiikan ehdoilla? (On Conditions of Water or the Metropolis Policy? Research report.)

Kolmisopimustoimikunnan esitys 9.1.1991. (Tripartite agreement. HSY Archives.)

Lindberg 24.4.2008. (Interview)

Water works annual reports 1970–1982

Vesi- ja viemärilaitostoiminnan yhteistyön kehittäminen pääkaupunkiseudulla (1988). Kolmisopimustoimikunnanehdotus ja mietintö. 18.11.1988. (Development of cooperation for water and sewage services in the metropolitan area. Tripartite agreement. HSY Archives.)

Vinnari E. (2008). Public service or Public Investment? An Assessment of the Consequences of New Public Management in the Water Sector. Doctoral Dissertation, TUT Publications 726, Tampere.

Newspapers

Helsingin Sanomat 7.5.1998.

Länsiväylä 26.9.1993.

Länsiväylä 5.3.1989.

Epilogue

Petri S. Juuti, Tapio S. Katko and Klaas Schwartz

There is more than enough water in the world for domestic purposes, for agriculture and for industry. The real problem lies elsewhere. The World Water Development Report 2003 already noted the major problem of the water and sanitation sector: "Sadly, the tragedy of the water crisis is not simply a result of lack of water but is, *essentially, one of poor water governance*." (UNESCO 2003). Unfortunately, that is still true today.

The chapters of this book have described a new approach to analysing water services provision. According to it, technological, socio-political and environmental dynamics (and their interactions) shape the provision of water services over time. We hope to show that the technological paradigm, though still prominent, is not the only relevant paradigm for understanding water services. For that purpose we present various frameworks for analysing water services over time. These frameworks, each emphasising a different dimension, highlight historical changes in technology, society and the environment in relation to water services. The underlying argument is that such a multi-dimensional approach is required to understand the current provision of water services and, more importantly, to address future challenges facing the sector.

As pointed out in the first volume of the "Governance and Management for Sustainable Water Systems Series" by IWA Publishing, water policy and politics are perhaps acknowledged by many water experts and professionals, but they are frequently overlooked and thought of as falling outside the realm of the water services sector (Grigg, 2011). In other words, despite being confronted with social, political and environmental issues and dynamics, few water professionals give sufficient attention to these phenomena. By ignoring these dimensions, possible alternative ways and strategies of addressing water challenges in professional activities are also overlooked. For a better understanding of water governance we obviously need to promote more inter- and multidisciplinary approaches than has been the case so far.

This book argues for the need and relevance of long-term thinking – understanding the pasts and presents to be able to identify probable, plausible and desirable futures. The book has four major sections: Water Governance Frameworks; Technology and Socio-Ecological Interaction; From Government to Governance. The 16 chapters from nearly as many countries and regions complemented by a few more global reviews show the diversity and complexity of water management and governance.

OVER-ARCHING THEMES

After reflecting on the research questions and frameworks presented in the prologue and the contributed chapters, the editors identified the following six major over-arching themes that seem to rise from these experiences: (i) Multi-level Governance, (ii) Speed of Change, (iii) Technopolitics, (iv) Increasing Complexity of Management, (v) How Good is Our Governance?, and (vi) Have We Really Learnt from History?

Multi-level governance

In line with globalisation processes and the growth of cities and population, water systems have expanded and are still expanding in many parts of the world. At the same time, people in smaller communities and dispersed areas often prefer modern small scale or on-site systems. Therefore, multi-level governance and management can be summarised as "*from local to global, and from global to local*". The role of *local governments* in water services is fundamental – something that has not perhaps been adequately stressed in global water forums and policy discussions. In developed countries local governments have overwhelmingly been the owners of water systems, and in transition and developing economies they are getting an increasingly important role through decentralisation.

In spite of lofty promises by the promoters of water privatisation, particularly from 1995 to 2005, it is now again clear that private ownership and funding are not the solution that will bring services to the people of the global village. It is a historical paradox that the policy which failed in the late 1800s has been reintroduced. Yet, privatisation is still largely promoted by international financial bodies drawing largely on public resources. While the private sector as producer of services is an important partner in long-term development of water systems, as it has always been, public responsibility for water services is a fundamental need of all communities that cannot be overstressed.

Speed of change

As argued by Kemerink *et al.* in Chapter 2, the pace and scope of change to the structures and dynamics of the state, in general, have become much greater than previously experienced (Thynne, 2000). At the same time, development of ICT infrastructure and a flourishing "global" water community have meant quick dissemination of new ideas and practices. As a result, water utility managers are increasingly being exposed to a highly dynamic environment in which they have to operate. Water utilities, for example, have abandoned hierarchical organisations and assigned more responsibilities to teams or operational units. These reforms, in turn, have demand professionals with a whole set of new abilities, such as working in teams and building consensus, critical thinking and problem solving, and communication.

Technopolitics

The concept of technopolitics introduced by Hecht (2001) seems to be quite relevant as shown by the cases from Spain, Kenya and Finland. Citizens' and even professionals' views can be bypassed by technopolitics- and metropolitics-oriented decision-makers. That may be justified in more business- and image-oriented sectors, but not in the field of services of necessity like water.

Increasing complexity

The operational environment of water utilities and services is becoming more complex. Not only do changes occur more rapidly, the governance structures have also become much more complex as a shift has taken place to "governance-beyond-the-state". This shift concerns the "emergence, proliferation and active encouragement of institutional arrangements of "governing", which give a much greater role in policy-making, administration and implementation to private economic actors, on the one hand, and parts of civil society, on the other, in self-managing what until recently was provided or organised by the national or local state" (Swyngedouw, 2005:1992). As a result, in most countries, public sector reforms have led to a "blurring of the frontiers between the public sector, the market sector and the voluntary sector" (Pollitt, 2003:28). This proliferation of agencies and actors involved in water management has greatly increased the need for co-ordination and negotiation skills. Utility managers need to take into account the different and often contradictory interests of various actors.

Since the number of water utilities in the Netherlands has decreased dramatically to ten, one could imagine that the environment would be less complex. Yet, due to the sub-companies and other activities, the whole arrangement has become ever more complicated and likely less transparent. If utilities become too large and powerful, regulation will be difficult. The number of utilities – like decision-makers may think – is not necessarily the key criterion, but rather how various organisations adapt to local natural and man-made conditions.

How good is our governance?

Decisions regarding water services have often been based on secondary criteria in a highly top-down manner without proper involvement of the parties concerned. For example, the free water policy introduced in several developing countries after gaining independence can be understood as a political reality of that time. However, such policies mainly benefitted local elites and eventually led to financially unviable utilities which only catered to a fragment of the population in their service area. Private sector participation was also highly promoted in the 1990s on ideological grounds. Little empirical evidence, however, existed to support the idea of the private sector being more efficient and effective than the public sector. Due to these decisions, many decades of developing water services provision have gone to waste. From the citizens' point of view, unrealistic policies blocked development during that period.

Governance is assessed below by using the key elements of Good Governance as defined by UNESCAP (2012):

As concerns the *Consensus/participatory/transparent* criterion presented in the Helsinki case, the professional experts' views were almost completely excluded when promoting technopolitics, a largely top-down and fashionable policy, which influences also other fields of Finnish society like education. In Latin America people seem to be more aware of their rights which, on the other hand, has been questioned. Indeed, participation of various stakeholders – from professionals to decision-makers, policy-makers, civil servants, citizens and other beneficiaries – in various forms, depending on conditions and traditions, should be promoted. Although not directly dealt with in the presented cases, the fight for more transparency against corruption has been placed on the agenda while a few decades ago it was considered too sensitive an issue – a positive development as such.

With regard to *Responsiveness and sustainability*, for some reason donor agencies are not very interested in the lessons learnt from earlier activities and not always sufficiently responsive to needs. As the cases from Kenya show that learning and policy development should be considered in a longer historical context, if poverty reduction and improved equity really are the goals of development co-operation. Water demand in Colorado is estimated to be higher than available resources by 2050 and will thus be a challenge for sustainability: the current patterns are to be improved and various alternatives explored.

Coherent integration of urban water systems, such as water supply and sewerage, is still in its early phases. Such integration has several advantages although in the case of inter-municipal and wholesale systems, separate organisations and operational areas can be justified. Less clear is the role of stormwater management which should be taken seriously due to climate change which alters rainfall patterns and intensity. The case from Kenya suggests establishing a national water council where all actors can deliberate and operationalise the water development agenda to promote the dynamism of water management.

As to the elements of *Equity and ethics*, the sector reforms in Kenya have resulted in improved economic performance, while service provision still appears highly unequal. There seem to be few incentives to improve services in the low-income areas since there is no monitoring of equity at the utility level.

Water services, including sanitation, should not been considered mere business they also have economic good. Public responsibility for these services and systems cannot be bypassed. Agreements, treaties and conventions have no value without ethical and moral commitments – an issue that may still be too sensitive to take up.

Have we really learnt?

Many, if not all, developed societies have a history of subsidising the establishment of rural water supply, some urban waterworks, and especially wastewater treatment and water pollution control. Now, almost in a historical vacuum, developed countries are promoting the ideas of full cost recovery and efficiency for developing economies. History is forgotten and cost recovery, which could be feasible as a long-term goal, is demanded by external forces and institutions.

Our cases and examples indicate that technological systems should not be seen as artefacts only, but also as systems and processes and related knowledge. One might also ask whether the artefact view of technology is still the major paradigm? As if technology existed in a historical or social vacuum.

New public management (NPM), the idea that public systems should behave like private ones, can be fundamentally questioned. Although full privatisation is not suitable for water services, NPM is promoted somewhat uncritically and without questioning its applicability. Water is fundamentally a natural resource and a basic human right (access to services), although it has also the role of an economic good, public good and cultural/social resource.

Many decisions and selections related to water management over time have created large numbers and levels of path dependencies. They are difficult to get rid of and limit available future options. Instead of unwise decisions, path dependencies should derive from sound and justified decisions that take into account long-term sustainability.

CONCLUSIONS

The following key conclusions can be drawn based on the contribution of the 16 chapters and our analysis:

(i) Some lessons have been learnt, although several of them are somewhat unclear and need to be explored further. Turning these lessons into actions and thinking about their implications for future policies are a big challenge.

(ii) The common time frames of projects, for example, in developing economies, are typically far too short. Instead of ad-hoc projects – that are probably needed as well – we should introduce the idea of continuous development and long-term thinking.

(iii) Current development policy and co-operation is based too much on ideologies and fashions. It is like a game of baseball where both the donors and key players try to catch the ball – symbolically competing about who is going to invent a term that becomes the latest fashion – instead of remembering the past lessons, practises, strategies and principles.

(iv) Sustainable water management and governance should be based more on the utilisation of accumulating knowledge or layers of knowledge.

On the whole, the obvious shift from government to governance is to be seen as a major challenge for the water sector and related professionals. How should we all act in the operational environment that becomes more diverse and complicated while the expectations of the public and citizens are getting higher?

As pointed out already in 2003 by UNESCAP, the removal of the constraints on water and sanitation service production and the inefficiency of sector organisations are essentially a *governance problem* in many countries. (UNESCAP 2003) Lack of *good governance* principles is one of the root causes of all major constraints within our societies. Good governance is participatory, consensus-oriented, accountable, transparent, responsive, effective and efficient, equitable and inclusive and follows the rule of law. It also ensures that corruption is minimised, the views of minorities are taken into account, and that the most vulnerable members of society are listened to in decision making. Furthermore, it is responsive to the future needs of society. In future, especially the needs of poor people have to be satisfied. The problems of the poor are expanding and their living conditions are weakening. The overview of the recent Human Development Report 2006 (UNDP 2006) summarises the issue as follows: The global crisis in water consigns large segments of humanity to lives of poverty, vulnerability and insecurity. The scarcity at the heart of the global water crisis is rooted in power, poverty and inequality, not in physical availability. There is more than enough water in the world for domestic purposes, for agriculture and for industry. The problem is that some people – notably the poor – are systematically excluded. This has to change in the future.

Finally, it is argued that better water governance requires that history and analysis of its lessons are taken more seriously than has been the case so far. The roles of various partners do change over time. Indeed, there is no end state in the futures and we have to develop water management and governance continuosly. Our actions will determine the development paths and governance principles of possible, plausible and desirable futures.

REFERENCES

Grigg N. S. (2011). Governance and Management for Sustainable Water Systems. IWA Publishing. London, New York.

Hecht G. (2001). Technology, Politics and National Identity in France. In: Technologies of Power. B. Michael, T. Allen, G. Hecht. (eds.), MIT Press, Cambridge/London.

Pollitt C. (2003). The Essential Public Manager, Maidenhead: Open University Press.

Thynne I. (2000). 'The state and governance: Issues and challenges in perspective'. *International Review of Administrative Sciences*, **66**, 227–240.

Swyngedouw E. (2005). 'Governance innovation and the citizen: the Janus face of governance-beyond-the-state'. *Urban Studies*, **42**(11), 1991–2006.

UN (2006). 2nd World Water Development Report. Water: a shared responsibility.

UNESCAP (2012). What is good governance? United Nations Economic and Social Commission for Asia and the Pacific. 4 p. http://www.unescap.org/pdd/prs/ProjectActivities/Ongoing/gg/governance.asp (11.06.2012).

Index

IWA Publishing's authorised EU representative for General Product Safety Regulations is Diane D'Arras, 15 rue Duret, 75116 Paris, France, e-mail: safety@iwap.co.uk.

Printed and bound by CPI Group (UK) Ltd, Croydon, CR0 4YY
12/05/2026
02108891-0011